Go to the table of contents

The World's Greatest Physical Science Textbook for Middle School Students in The Known Universe and Beyond!
Volume One

The scientific method

Matter

Energy

By Michael A. Ritts

February 2016

Note:

Volume Two – Motion, forces, and physics

Volume three – Chemistry, waves, and pseudoscience

Copy write © 2016

Super George Productions

Ethics and honesty

This book is meant to be a textbook, treat it as one. If you buy a personal copy you can lend it just like any book, but do not copy it. If you are a School and you have 500 students per year you should pay for 500 copies, but you should not have to buy it again the next year, or the next ten. Just like any hardback text, you buy it once, and are good for quite some time. You get your textbook and I get my money. Honesty is a great thing.

The Internet and computers are wonderful things, but full of copyright infringement. Do not lower yourself because you can get "free stuff". Someone is getting ripped off, whether it is a writer, singer, or movie company. A lot of time, money and talent go into these products, and those people deserve to be compensated. As the cost of products, in the form of profits, goes down, so do the rewards to these hard working people, bad for everyone. Everyone wants "something for free", but when you think about it, it does not pay off in the long run. The day music is free, no one will make **new** music, the day medicine is "free" no one will make **new** medicine, the day textbooks are free, no one will write **new** textbooks.

So please, be honest, it is an honorable thing to be.

Table of contents – Volume one

Why does this book exist?

How to become smarter

How to become smarter the interesting way

Link to all my science videos: https://www.youtube.com/channel/UCNERGz70pAoK8A5oe2Wkw_Q

This is <u>volume one</u> of a three volume set

Why does this book exist?

In a nutshell, efficiency and cost. This is meant to be a "new age" middle school textbook, organized in such a way that the teacher can follow it chronologically, in a sensible pattern. It is written immaturely, so that middle school students will actually read it. It is not watered down, the vocabulary is there, but admittedly it looks like it was written by a 13 year old. That was my goal. With the Internet and this eBook I have the ability to include links to videos, simulations, demonstrations, and anything I believe will make the book more interesting. These are things a physical book cannot do. This book is intended to be an eBook.

With school budgets in crisis, and current textbooks being very expensive, this book is made to be as affordable as practical. My competition are the major textbook companies who charge in the neighborhood of $100 per text, granted they come with a lot of junk, worksheets, test, videos, etc. But to tell you the truth, these are not really very useful. For a school of 500 students, that comes to quite a chunk of change. How about if we could cut that cost by a factor of 10? The biggest selling point of the old textbook my district spent a ton of money on was a free stick drive for every teacher. Really that is why we chose it, a bribe. On top of that, all the different middle school textbooks are nearly the same. Not much choice out there.

Current textbooks are huge; they contain *way* too much stuff. They are *not fun* to read. They were written by intelligent people, who really do not understand the middle school student's mind. I am a teacher, I know how they think, and I am not sitting in an office somewhere imagining it. I may not be as smart as those writers, but I do not have to be, I only need to accurate and interesting. This is what I have tried to do. Granted there are some "free" textbooks available online but they are rather out dated and not "aimed" at the middle school student. This one is. There may be some mistakes and if you find something you cannot tolerate, e-mail me at junglecat3388@gmail.com and I will fix it (if I agree).

So what do we have here? An inexpensive e-book textbook that is easy to use and fun to read. Time will tell…….

So what am I visualizing here?

I hope to make the price affordable for any student or school district. It took two years to write, I should get something. I am also saving up for a new ping-pong ball, so I need the money.

You will find many links to videos of some of the demonstrations I do in my class, along with other on-line resources (simulations and things).

I wrote this book exactly like I teach the lessons. The stories and examples are the same ones I use in class, so I know they work.

Textbook Organization
Each chapter is organized into four parts.
1. A link to a video of the actual power point lesson I use in class. It is not the same as an actual lesson but it is a good review or introduction. It will also give the teacher some good ideas.
2. An anecdote about science – it may not have anything to do with the chapter's topic, it is not meant to, it is to try and spark interest and curiosity.

3. The Body – This is the information part – accurate but full of silly examples to help explain the topic. **There is a link to <u>flash cards</u> at the end of each chapter**.

4. A day in the life of Earwig Hickson III – the creative part. Earwig is an imaginary child full of wonder, curiosity, and imagination, like we all should be. Since he is imaginary (think a Saturday morning cartoon character) he can do things that really are not possible. His imagination can run wild without any dangerous consequence.

How to use this book

To the teacher

It is a textbook, an outline, and a teaching strategy. You as the teacher still have to present the material, come up with lesson plans, write your own tests and review exercises. A good teacher is not lazy; they just need a foundation to work with.

To the student

Read it, check out the web links and watch the videos. Find out that science is fun. You will be smarter after reading the book that is something to be proud of. I hope to teach you, if nothing else, that science is fun, I mean awesome. Science rules!!!!!

How to become smarter the interesting way

With the Internet and especially YouTube and Wikipedia, interesting information is everywhere. Since I am a scientist, I have my heroes. This is a list of my role models. They are special to me, not because they are smart or made great discovers, but because they are interesting to read about, some are super funny, some explain complex things in simple language, and some are not even scientists.

These are some of the *giants* I learned from

Bill Nye – Google any *Bill Nye the Science guy* topic on YouTube and learn a lot. Some people think Bill is just an actor on a T.V. show, but he is more than that. He is a real scientist who just happens to have a talent at being funny, and he explains things so anyone can understand them. He even designed an experiment that was put on one of the Mars Landers, quite an achievement. He has probably done more for science in recent years than any other person. I would love to meet him someday. He does not know who the heck I am. http://billnye.com/#educational

Neil deGrasse Tyson – He is the current "Rock Star" Scientist. He is on T.V., radio, and all over the Internet. He is smart and very funny. I love his quotes. He is currently the director of the Hayden Planetarium at the Museum of American Museum of Natural History (the one the movie *Night at the Museum* took place in). His language can be a bit salty sometimes, but you are mature and can ignore that. His life story is interesting. He is awesome. He does not know who I am either. http://www.haydenplanetarium.org/tyson/

MythBusters – The TV show follows the scientific method extremely well. The topics they investigate are interesting. They do an excellent job and are very fun to watch. They have investigated thousands of interesting myths. They always follow the scientific method, they are true scientist, but will probably never admit it. They do not know who I am either. Their web site has tons of interesting video clips. http://www.discovery.com/tv-shows/mythbusters/?rfc=1

Ted-ed – an organization that produces excellent short videos about interesting things. http://ed.ted.com/lessons?category=science-technology

The Big Bang Theory – This T.V. show is not only funny, but also accurate. They actually have a scientist check the script for accuracy, a rarity on Television. By the way not all scientists are like Sheldon, just some. They do not know who I am.

The skeptic's dictionary – sometimes it is hard to figure out what is real science and what is not. This site will help clarify that. The topics are fun to explore but often the "proof" is disappointing. I personally like reading about Sasquatch and flying saucers for entertainment, but not for knowledge. They do not know who I am. http://www.skepdic.com/

Albert Einstein – One of the more famous scientists in recent history. He was the first "Rock Star" scientist, which means his name is recognized by nearly everyone. He was an interesting person, and a very smart one. He made some interesting quotes. He has absolutely no idea who I am.

Sir Isaac Newton – you probably have heard of him. He is arguably the most brilliant scientist ever. Other scientists since his time are smarter, but they had *his* discoveries to work from, he had very little to start with. He came up with his discoveries almost from scratch. He was, in my opinion the first great scientist, and probably the smartest man who ever lived, although I do admit there were great ones before him. He may not be as exciting as some people, but smart people do deserve to be remembered. He does not know who I am either, mostly because he died 300 years before I was born.

Carl Sagan – Probably the second "Rock Star" scientist. He developed a super interesting T.V. show called *Cosmos*, which was educational and easy to understand. He was very interested in finding alien (not of earth) life on other planets. Sadly he died before we found any (by the way we have not found any YET). He wrote some excellent books, by far my favorite, is called **Demon-Haunted World: Science as a Candle in the Dark**. This book explains the difference between science and non-science (pseudo-science) and is held in great esteem by me, mostly because it was the first time I learned about *sleep paralysis*, which affected me as a teenager and caused me great concern. By the way if you think aliens are visiting you in the night, or demons or anything else, Google *sleep paralysis*, it may explain much. Carl did much to popularize science, a great man. He has no idea who I am.

Steve Spangler – An ex-science teacher who spends his time teaching other science teachers and motivating younger kids about the wonders of science. His specialty is very cool demonstrations that amaze and educate. He also has a store where you can get safe-cool experiments. He does not know who I am either. Check out his website at http://www.stevespanglerscience.com/.

Richard Dawkins – A very good biologist. He wrote the first book that helped me understand how natural selection and evolution worked called **The Selfish Gene: 30th Anniversary edition**, a great book. Unfortunately he has recently become embroiled in a battle with non-scientific thinking kind of people and they have tried to damage his reputation. Some people really hate him, but trust me, you can learn a lot from him. I respect him. He does not know who I am either; I am beginning to think I am a real loser.

Spend some time surfing the Internet for these guys, and you will be glad. They are true heroes. I will say this many times, smart people deserve to be Googled too!

Unit 1
What the heck is science?

Chapter 1
How to think like a scientist

Video of the actual power point lesson:
https://www.youtube.com/watch?v=K9j1rZt41_I

The wonders of science

There is no such thing as *magic* only science. Mankind's greatest discovery was not sliced bread or even the wheel. It was science, or more specifically the *scientific method*. This changed the world and is still changing it today. It is sometimes called the scientific revolution, but whatever you call it, the world changed in a series of steps like a row of falling dominoes. Some say it began in the 1500's, but it had been slowly gaining momentum long before then. Out went the dark ages and in came the age of enlightenment. Advancements started coming faster and faster. So many things changed resulting in a new world, one of understanding, one of problem solving, one of discovery, one of making life better for everyone.

You owe a lot to the scientific method, even if you do not realize it. Safer food, more food, better medicine, vaccines, transportation, communications, electricity, higher life expectancy, clean water, computers, cell phones, antibiotics, and genetic engineering. The list goes on and on. Chances are you or someone you know is alive because someone somewhere followed the scientific method.

The story of vaccines, https://en.wikipedia.org/wiki/Vaccine

The scientific revolution, https://en.wikipedia.org/wiki/Scientific_revolution

How corn (maize) fed the world, https://en.wikipedia.org/wiki/Maize

How to think like a scientist

You used to think like a scientist, what happened?

Thinking like a scientist means thinking like you did when you were younger, like two years old, remember that? When you were a small child you were curious and got into everything. That is what scientists do, they get into everything. Some of the best scientists were terrors as children and drove their parents crazy. Those that did not get punished to have their curiosity driven out of them became scientists. So back to your childhood we go.

So what was it like back then? Were you skeptical of everything your parents and other adults told you? Did you want proof of their claims? Did you like to take things apart to see how they worked? Did you burn your hand on the stove looking for the source of food you knew came from there? This is what science is.

Remember when you first suspected that your parents lied to you about Santa Claus? Did you suspect something? An old man who flew around the world with magic reindeer to give every good kid presents. Did you realize that the all-seeing Santa missed some of the things you did? Did you notice your sister always got presents and you knew she was, in fact, pure rottenness? Did you try to report her in to the elves? Did this peak your sense of wonder. Did you suspect it was all a lie to make you behave? My kid tried to set a trap for Santa. I messed the trap up so he would think Santa escaped. I am guilty too. Santa is fun.

A scientist is **skeptical**; they do not believe anything without proof (or data). They rely on **experimental data** when making a decision. They value experiments that have, or can be, repeated by anyone. Scientists like to make **predictions** and then test them to see if they come true. If an experiment does not work out the way they think it should, they adapt to new observations or new experimental evidence. All scientific experiments are **verified** by other scientists to ensure honesty and accuracy.

I remember one of my first experiments. My Sunday school teacher told me that on Christmas morning, and I mean 12:00 midnight, that animals could talk. It was a Christmas miracle she said, and I believed her. I guess she never thought I would test it. I wanted to talk to my pet cat. She did not speak. I never trusted that lady again. Her claim could be tested, a prediction could be made, the experiment done, and it failed.

The hypothesis

What is a hypothesis? Is it an educated guess? Not really, it is much more than that. When you make a guess, you really are not acting like a scientist. A good hypothesis makes **a prediction**, one that can be **tested** with an experiment. In fact that is exactly what a hypothesis is: a prediction that can be tested with an experiment.

OK, maybe a hypothesis is an educated guess but only in the way that it has no proof, it is a gut feeling. You think it might be true, but you have no data to say it actually is. The other cool thing about a **good hypothesis** is that no matter what, it is not wrong, at least not at first, because a hypothesis has to be **proven** wrong. Once it is proven wrong though, then it is always wrong.

So what is a **good** hypothesis? It is a testable statement that can be proven or disproven with an experiment. **It is an IF, THEN statement**. It is a prediction. It must be testable with an experiment.

Bad hypothesis (not a hypothesis)
I think Santa Claus is real.

Good hypothesis (is a hypothesis)
If I stay up all night, **then** I will see Santa.

See the difference? The second one can be tested with an experiment, stay up and see. The first is just an opinion and cannot be tested with an experiment.

So what exactly is a theory?

So what is a **theory**? Why it is one of the most misused words in the English language. People like to use the word theory when they actually mean *opinion*. I guess they think it makes them sound smarter, but, in science we do not try and sound smarter; we try to be right or at least useful to society.

A theory is a specific scientific term that means a specific thing. It is *not* an opinion, as some people seem to think. When a scientist says they have a theory, it means something. It means there is a lot of data to back it up. And I mean *a lot*; we do not use the word theory lightly. A good theory is the **absolute best explanation** to describe observable facts. But that is not all; a good theory also makes **predictions** that come true. These are called **valid predictions** as opposed to guessing. A good theory has a lot of data to support it. It is not a gut feeling or common opinion or belief. It has data and a *lot of it*. It has been tested and **passed every experiment** with flying colors. The absolute most important thing about a theory is that it makes predictions that you would actually bet your **own money** on. But this does not mean that a theory is a fact, it might not be, but it is the *best* explanation for observations and facts. The bottom line though is that if scientists have a theory, you can take that to the bank, it is the *best we have*.

Some of the theories we laugh at today were great theories for the time. The theory that the earth was flat proved false, but it was a useful theory, because it explained **all** observations and kept people alive 4000 years ago. Back then if someone sailed an itty-bitty boat (navigation had not been perfected yet) out of sight of land, they risked death from getting lost or killed by a storm out of sight of land. Those that believed that theory lived, and those that did not, may have never came back. By the way in Christopher Columbus' time the Earth was known to be round, he did not "discover" the earth was round as some people have been taught.

Some people use the word "theory" when they mean, "believe". It may make their argument sound better but, in fact they are lying. When I say I have a "theory" that Bigfoot exists, where is my evidence? If I use the word **theory**, it is fair for you to ask for that evidence. The fact is, without data or experimental evidence, it is just my *opinion* and not at all scientific.

A good theory

A theory about how gravity behaves (actually a **law** but we will get to that in a minute) is a great theory because it predicts what will happen if you jump off a cliff. There is a lot of evidence to support it. Everything we ever dropped over a cliff went splat. It makes predictions that come true *every time*. Only a fool would bet against the theory of gravity. By the way I have no idea HOW gravity smooshes me when I jump off a cliff, only that it does, and I wish to live.

A bad (or non-theory)

"I have a theory that if I believe strongly enough I can fly." This is a horrible idea because there is no data to support it and it cannot be tested, and live, and if you were to try and test it by jumping off a cliff and flapping your arms, bad things will happen. At best it could be called an opinion but is still a bad idea. *Everything is a bad idea without scientific evidence.*

it seemed like such a good idea at the time

COMPARE AND CONTRAST

HYPOTHESIS

NOT PROVEN

TENTATIVE
EXPLANATION
IF - THEN

BEFORE THE
EXPERIMENT

SCIENTIFIC

ANSWERS
QUESTIONS

THEORY

HAS DATA TO
SUPPORT IT

BEST
EXPLANATION

AFTER THE
EXPERIMENT

MAKES PREDICTIONS

So is a law true?

For the most part a **scientific law** is considered true by nearly all scientists, at least it has never been disproven, but it is not a theory. A *theory* explains **why** observations occur; a *law* only describes **what** happens. By the way even a law is not considered "absolutely true" they can be modified, but they are considered nearly complete.

The **Universal Law of Gravity** basically says things denser than air fall to the earth when dropped. It has always happened that way and it always will. It actually says a lot more than that, but I will save that for later. No sense getting flooded with too much information too soon, one step at a time.

The current theory of gravity (and there is more than one) tries to explain **why** things fall to the earth. So far we do not really know for sure why things fall to the earth. Newton was stumped, Einstein gave it his best shot, and currently scientists are coming up with new ideas.

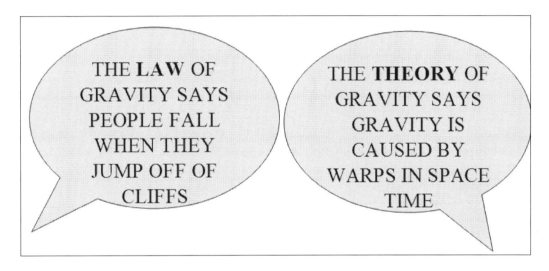

How a Law Is Developed
In a nutshell this is a simplified explanation of how the **Universal Law of Gravity** was formed. A brief history lesson!

*It is said that sometime in the 1500's an Italian chap named **Galileo Galilei** had a hypothesis that all objects, no matter the mass, <u>fall at the same rate</u> on Earth, when ignoring air friction. He tested it and found it to be true. This went against common sense at the time because people were used to the invisible air causing lighter objects to lag behind. Imagine a feather falling at the same rate as a hammer, but that is what happens if air is not in the way, to mess it up. This experiment was actually done on the moon, but scientists were rather sure it was going to work. They did it anyway because that is what scientists do. http://er.jsc.nasa.gov/seh/feather.html*

*After similar experiments were conducted by others over many years and different situations, the idea was considered verified. At this point it could be considered a **theory** because it had a lot of experimental data to support it and it made predictions that came true.*

*Then in 1687, <u>Sir Isaac Newton</u> built on the idea and discovered that all objects with **mass** attract all other objects with mass at the same rate. He even came up with some mathematical formulas, which predicted this attraction. His conclusions were so complete and explained so many observations (including the motion of the planets) that it could be considered a law, <u>**Newton's law of Universal Gravitation**</u>. The predictions were so accurate that 300 years later NASA used his math formulas to land people on the moon (a very difficult feat by the way). His law explained exactly **what** objects do in a gravity field but **not why**.*

*The Law of Universal Gravitation was not quite perfect though, it took <u>Albert Einstein</u> to perfect it in 1916. He tried to explain where gravity comes from; he called this The **Theory** <u>of General Relativity</u>. This theory is still being tested and verified today.*

And then there are opinions

<u>Opinions</u> are beliefs. These are the backbone of our personalities. **Anyone can have *any opinion they wish as long as they do not call it science*.** If I have an opinion that aliens are abducting people and peeling their skin off, that is OK, but only if I say it is my **opinion**. If I call it a theory I am lying because there is no data to support it.

Many people use the word theory when they should say opinion; *I have a problem with this*, at least be honest enough to say it is an opinion, don't try and pawn it off as science when it is not.

The other thing about opinions is that they cannot be disproven. No matter how many times you do **not** see Santa Claus, it does not mean he does not exist. This is called proving a negative or <u>evidence of absence</u>, and you cannot do that.

I am a hunter; I have not seen a deer in 20 years. Does that mean that deer do not exist? I might think so, but that would be my *opinion* because I have no data. Someone who wishes to prove deer **do** exist only needs to catch one and show me. How could I prove they do not exist, by not catching one? How could someone prove Bigfoot does not exist, or the Loch Ness monster, flying saucers, or the flying spaghetti monster. You can't. *But they would all be easy to prove*, just catch one.

The word theory used correctly

The theory of <u>evolution</u> (or <u>natural selection</u>) is a good theory because it explains all observations and makes *valid predictions*. One prediction that was made (back in the 1970's) was that disease-causing bacteria would become immune to current antibiotic medicine. This turned out to be true and sadly there are now diseases that we cannot cure (although we could 40 years ago). As the bacteria evolved they have become immune to more and more of our antibiotics. These bacteria are often called "bullet proof" bacteria or more properly <u>MRSA infections</u>, which are immune to our current antibiotics, a very bad thing. Perhaps it would have been best to take this theory more seriously.

Here is a fun game which shows how natural selection works.
https://phet.colorado.edu/en/simulation/natural-selection

The word theory used incorrectly

The "theory" of creationism is *not a scientific theory*; it is an *opinion* or belief about how the world was formed. There is no data to support it; it makes no predictions that come true. It does not explain observations. It is not science. **It is OK to believe it though, but at least admit that it is your <u>opinion </u>not a theory**. There are people who believe an opinion so strongly that they insist that since they think it is true, others must believe it too. This is when they try and *force* it into science. Creationism or *Intelligent design*, as it is now called is such an idea. Using *bad* science and *political power* they have lately tried to *force* it to be taught in public science classrooms. This is wrong, not because I do not believe it, but because there is **no scientific evidence** to support it. If you are thinking as a scientist, right now, you will now start your own research. <u>What do scientists think about this?</u> https://www.youtube.com/watch?v=b_R5tv9LLGo, https://www.youtube.com/watch?v=VO969i-qgxE

By the way, if you have an *opinion* it is OK to believe it, but *do not call it science* and do not bet money or your life on it, and most of all *do not force it on another*. This is just not nice.

THEORY		BELIEF
ALWAYS HAS DATA	IDEA	NOT FOUNDED ON PROOF
MUST BE TESTED	ANSWERS QUESTIONS	DOESN'T HAVE TO BE TESTED
MAKES PREDICTIONS		OPINION
EXPLAINS FACTS		CAN NOT BE DISPROVEN
VERIFIED CAN BE PROVEN WRONG		

OK, so how do I think like a scientist?

You already do think like a scientist, I know because you are still alive. Think of what your great ancestors were doing 15,000 years ago. Those that thought scientifically lived to have children, those that did not, died, and dead people do not reproduce.

A story about survival or not

Long ago, we will say 15,000 years ago; two friends were walking along in the savanna of Africa. One was your great ancestor and the other was named Bob. A hungry saber-toothed cat jumped out from behind a rock. Your ancestor ran away, but Bob said "here kitty kitty". Bob had no children (because he was dead) but your ancestor did. You are here and Bob's descendants are not. How sad.

When you think like a scientist you survive. The first thing you do is **observe** your environment, you use your senses. When you see something interesting or threatening you **analyze** it, or think things through. Do you have any experience with this; is it dangerous, can it be eaten? You then **synthesize** a hypothesis. This is your experiment. **IF** I stay, **THEN** I will be eaten. Finally you test your hypothesis. If you thought it was dangerous you ran, if you thought it was edible you ate it. You must have good scientific genes because you are still alive and Bob's descendants are not.

A day in the life of Earwig Hickson III

Earwig is our all-purpose child-scientist, many of the things he does you can **only do once**, but since he is imaginary it is OK.

Earwig noticed a magic box in his house. When someone put a shiny metal disk into the box wonderful things happened. He *observed* how the magic box made the shiny disk better. The shiny disk was turned into an amazing movie about airplanes! Immediately his mind went into *analyzing* his observation. He reasoned that anything that goes into the magic box would be improved and he knew of a lot of good things the box might make better. Dozens of ideas popped into his head about what he could put into the box. Waffles were good already but how would the box improve them?

This is when he *synthesized* an idea. What would happen if he put a waffle in the magic box? He did not know it at the time but he just formed a *hypothesis*. **If** he put a waffle in the magic box **then** something wonderful would happen. As wonderful as this discovery sounds, he did not discover anything yet. He had to test it. It is *not that he is a troublemaker or a bad person, he is a scientist* – he had to try it. So he did. In the waffle went, with a little force he got the door closed. Nothing happened. His experiment failed. But did it? Not really, *a failed experiment gives you knowledge you did not have before.* **Failure is success**! Now he knows that waffles do not go into the magic box. He got grounded. All *Scientists fail **all** the time, but in the end they get it right.*

He learned from the failed experiment and came up with a new and improved hypothesis. Since the disk had a hole in the center maybe a doughnut would work better. In the box the doughnut went, he got grounded again. His parents were frustrated and worried. *He was not a bad child, he was a scientist.*

The term "mad scientist" usually refers to the scientist; actually it is the parents that get mad at the little scientist. Earwig ended the investigation in the corner pondering the meaning of the sentence, *"If you ever put a jelly doughnut into my DVD player again I will whip the living tar out of you"*!!!!! What did this mean? Do people have tar in them? Does whipping get it out? Does this apply to trying a pita pocket or fruit roll up? And what is a DVD player anyway? So many hypotheses to test and experiments to do.

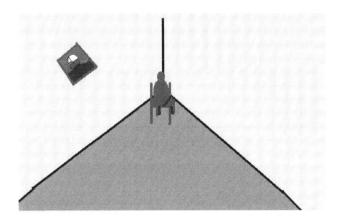

Review of terms – Quizlet: https://Quizlet.com/124227272/chapter-1-volume-1-how-to-think-like-a-scientists-flash-cards/?new

Fun things to Google

Newton's Universal Law of Gravity
Sir Isaac Newton
Theory of relativity
MRSA - *Methicillin-resistant Staphylococcus aureus*

Links

These are worth looking at, many are demonstrations I use, that make you think or better yet amaze you. Science!

How simple ideas lead to scientific discoveries - Adam Savage (from Mythbusters) explains how a scientist is made. http://ed.ted.com/lessons/how-simple-ideas-lead-to-scientific-discoveries

Skepdic.com – find out what is real and what is not: http://www.skepdic.com/

Galileo Galilei: https://en.wikipedia.org/wiki/Galileo_Galilei

Theory vs. opinion – A discussion about the difference between the two. https://www.youtube.com/watch?v=ekgzdTIiE8E

Not all scientific studies are equal. Why is it important to be skeptical? Ted-ed. http://ed.ted.com/lessons/not-all-scientific-studies-are-created-equal-david-h-schwartz

What is the danger of science denial? Are vaccines bad? Are alternative medicines good? Ignoring science can kill you! Is genetically modified food bad? A film from Ted-ed. http://ed.ted.com/lessons/the-danger-of-science-denial-michael-specter

An experiment on gravity hill – where things roll uphill! – Every area has a *gravity hill*. There are myths that go with this, but has it ever been tested. Yes. I tested it! https://www.youtube.com/watch?v=bsMZkm66uSI

Science vs. pseudo-science – Presidents Washington and Lincoln explaining the difference https://www.youtube.com/watch?v=c8W-ukldHmo

Newton explaining his Universal Law of Gravitation part 1 - https://www.youtube.com/watch?v=kgtfRTKYIsw

Newton explaining his Universal Law of Gravitation - part 2 https://www.youtube.com/watch?v=sUVUGng05iQ

Celts – a cool little object with a spin bias, a great Christmas present. If you want one, Google *Rattleback*.
https://www.youtube.com/watch?v=ps2KfAtYmCE

Pendulum springs – this is just plain cool. Can you predict what would happen, I can't.
https://www.youtube.com/watch?v=MVw6wiJqH6Y

Shapes on a balloon – Can you make a prediction? Do the results of the first test hold for the next?
https://www.youtube.com/watch?v=5vTQsSRNlKM

A center of gravity "trick" you can make with two forks, a cork and a needle.
https://www.youtube.com/watch?v=Gf_ofCPGk6I

How does a tight ropewalker stay balanced? This is a "model" that shows how.
https://www.youtube.com/watch?v=Jjg4rxlrQRs

This is puzzle – do you think you can make one?
https://www.youtube.com/watch?v=EVB70aeGZ_0

Another trick – why does the pendulum act like this?
https://www.youtube.com/watch?v=vVvKW5ok5nI

A cup of water that never runs out? Is it magic or something else?
https://www.youtube.com/watch?v=Pe8rJQkquBU

A running faucet appears to float in the air.
https://www.youtube.com/watch?v=kfRpSACuJk4

Chapter 2
The scientific Method

Movie of this power point lesson:
https://www.youtube.com/watch?v=wDpxjZb3Nd8

The wonders of science

It happened somewhere in our past. At least a few hundred thousand years ago, maybe as far back as 1.5 million years ago, but someone followed the scientific method and learned to control fire. It probably happened many times, but it changed our species. It made us different from all other life on earth. The great apes didn't do it, neither did the dolphins. Humans did and it made us the only advanced species on Earth. Fire eventually led to larger brains (cooked meat is more nutritious than raw meat), protection from predators, and even our beautiful facial features. Fire was the first step to what we are. I imagine the early cave people thought of it as magic, but it was not, it was science.

Where would you be if some caveman did not figure this out? Not where you are today I imagine.

The history of how fire changed humans,
https://en.wikipedia.org/wiki/Control_of_fire_by_early_humans

A fire bow making smoke: https://www.youtube.com/watch?v=APCgA9uxMGI

Learning through inquiry

Do not fear, do not be afraid. The scientific method is not scary. It is not even a useless list of vocabulary words designed to *suck the life* out of science class. It is just an idea. It is common sense. You use it all the time, but you did not know it. Back to your childhood we go, back to when you were two years old and thought like a scientist.

What you did back then was ask yourself and others questions. Why is the sky blue? Why do rocks sink in water? Why do dogs sniff dog dirt? This is called **inquiry** and simply means to learn by asking questions. This is what scientists do, they ask questions then try and find the answers.

The **scientific method** is an *idea*, a way of organizing your thinking, a way to develop a *strategy* for solving a problem. It is valuable because it helps *organize your thoughts;* it gives you the framework for a plan, a plan that will work every time. It is a thought process, but a thorough process.

Unfortunately we have to name the steps in this process and this is where we lose people. Uggghhh – vocabulary. I know this is what sucks the life out of an experiment, especially when we use the word *lab report*. The reality is the scientific method happens *in your brain*. The lab report is something to report to others.

The yucky part first – the steps to the scientific method

There are many versions of the scientific method, they are all the same idea, but have more or less steps, and different vocabulary. But remember the scientific method is really *just an idea or strategy*. It does not matter which version you use, just that you use one. I will be using a 6-step version. These are the steps.

Problem or question
Research the problem
Form a Hypothesis
Collect data – do an experiment
Analyze the results of the experiment
Conclude – answer the original question

It is helpful to remember the order of these steps and since my brain is small I use a sentence to help. **P**aul **R**eally **H**ates **C**hips **A**nd **C**hicken, **P** is for problem, **R** for research, **H** for hypothesis, **C** for collect data, **A** for analyze, and **C** for conclude. It just makes things easier. By the way you can make up any sentence that works, the more perverted the better you will remember it.

At this point (if you are normal) you are ready to quit reading and play computer games, but don't, the boring part is over. Once you realize what this means, you will discover you do this all the time, you did not call it the scientific method. You called it common sense. Let me show you.

Problem or Question

The first thing you are supposed to do is "*State the problem or question*" you are trying to solve. This can mean different things in different situations, but in real life you do it all the time. You may not "*state*" the question out loud but you do in your brain.

Sample problems
I have to go to the bathroom and my sister is in there doing her make up.
My feet hurt at the end of every day.
I have a headache.
A car is coming towards me.
I over slept again and can't afford to be late for school.

Sample questions
How do I get my sister out of the bathroom?
How do I get my teacher to stop harassing me?
How do I get a raise in my allowance?
What will happen if I mix kool-aid and bacon?
How can I cure cancer?
How can I avoid getting punished for putting a jelly doughnut in my dad's DVD player?

These are all questions that can be solved by using the *scientific method.* You probably already solved some of them.

Research the problem

Students do not like the word *research*. In the old days this meant a trip to the library for a long day. It can still mean that, but with today's technology it is a piece of cake. Research is not really looking for answers, you are looking for a *plan,* or hints about where to start. The Internet is today's library. You will find both good and bad ideas. You do need to sort them out, find verification by using multiple web sites. Some sites are better than others, learn to recognize them. Ask people or use your past experiences. Talking to someone who knows more than you do about a topic is research. Again, you are *not looking for answers*, just a strategy or a place to start. Brainstorming in a group is always a good idea.

Form a Hypothesis

This is the most important part and the hardest for some people, but if you remember to start your idea with the word IF, things get easier. Remember a <u>hypothesis</u> is your best idea written as an IF-THEN statement. It makes a prediction of some kind. It **must** be testable.

If you did good research, you should have some good ideas, if you don't, do more research. Now just take your idea and make a prediction, one that is easy to prove right or wrong. It is worth noting that no hypothesis is wrong at this point, as long as it is stated properly.

Good hypothesis

If I complement my mom's hair, **then** she will not punish me for as long.
If I ask politely, **then** my sister will get out of the bathroom.
If I stay up later, **then** I will get up earlier.
If I stop sleeping in class, **then** the teacher will stop bugging me.
If I move my alarm clock to the other side of the room, **then** I will get up earlier.
If I do my chores before being told, **then** I will get a bigger allowance.
If I jump off a cliff and flap my arms, **then** I will fly.
If I bring a rabbit foot to class, **then** I will get an "A" on my test.
If I throw a tantrum, **then** I will get a cookie.

These are all good hypothesis not because they are good ideas (some of them are not) but because they can be tested and proven right or wrong. The best may survive; the bad ones will die rather quickly. Sometimes you will have more than one good idea and thus more than one good hypothesis. This means you have to do more than one experiment because an experiment can only test **one** hypothesis at a time. Of course you could run several experiments at the same time, to test several hypotheses, if you do this make sure you take good data so you remember what you did.

Collecting data

This is the experiment, the fun part (usually). In this step you come up with some way to test your chosen hypothesis. Remember, you can only test one hypothesis at a time. *This is important.* An experiment can be a complex long project or can be simple. As long as it proves the hypothesis right or wrong it is a good experiment. If you set up a good experiment you will always learn something, so a failed hypothesis is not a failed experiment. Thomas Edison is said to have failed to make a light bulb 10,000 times. These were 10,000 failed hypotheses but since he learned 10,000 ways **not** to make a light bulb, they were successful experiments, just disappointing. In fact the most common thing a scientist does is FAIL! I may fail 10,000 times but in the end I will get it right, just like Edison. The key is getting it right in the end, scientists do not give up, they *fail until they succeed*!

You have done many experiments in your life that failed. Remember the first time you tried to steal food off the stove and burned your hand – you learned something. When you threw your first temper tantrum, you learned something – either you could manipulate your parents or you could not. Maybe you ate something brown you found in the back yard – you learned something. Did you ever change the TV channel with 2 minutes left in a football game your dad was watching – not all experiments end well – but you learn something.

So far I have not mentioned actually collecting data yet. In a way I have, we just did not write anything down. So what do you write down? The results of the experiment. A lot of times scientists like to use numbers, this is because numbers are easy to compare. Sometimes yes or no is good enough to describe the results. Edison timed how long his light bulbs stayed lit so he could compare different versions easily. A bulb that only lasted 10 minutes was a fail, but lasted longer than the one that burned out after 1 minute, so maybe it had potential (with modifications) after all. When you tested the food from the stove hypothesis that was more of a yes or no answer.

The best data and easiest to make and read are charts, tables or graphs. Now you know why you need to study Math, so you can learn science better!

We scientists have a saying, "Screwing around and writing it down is science". This means if you do not take data, you are just playing.

Analyzing the data

This is the simple part. It just answers the hypothesis. Did it work or is it a fail. Are you going to try to put your hand on the stove again? How did Edison know which light bulb filament to choose? How did the brown stuff you found taste? How did your parents react to your temper tantrum? What did your dad do when you turned off the football game? You looked and analyzed the results, the data.

Conclusion

The conclusion is easy too. You just answer the *original question or problem*. Did you find a way to get your sister out of the bathroom? Did you manage to avoid punishment? Did you get a raise in your allowance? Did you cure cancer?

Using the scientific method to get your sister out of the bathroom

Problem:
Your sister spends way too much time in the bathroom and you need to go bad, and you do not want to be late to school, because you have science first. You ask yourself; *how can I get my sister out of the bathroom quicker?*

Research:
You hit the Internet and start surfing. You find some pages where people posted similar questions. You find pages that explain just what girls do when are in the bathroom (makeup) and why it takes so long. You find a video where someone put a rubber snake in his sister's underwear drawer as a joke. The results were hilarious. Could this be applied to your situation? It is not an answer but now you have a *diabolical plan*. What could you do to get her out of the bathroom? Nothing that will harm her that is for sure, but something effective. Would loud music she hates drive her out? (Probably not since she is playing loud music anyway) Maybe bad smelling gas under the door (no you need to go in there too). How about a spider (you reject that because you hate them more). Maybe a rubber snake under the door. Wait – what about a real snake!!! *There is a plan.* You already have a pet snake, turtle and a frog. Which would freak her out the most? You come up with 4 hypotheses. You test them all.

Hypothesis:
If I slide a rubber snake under the door, **then** she will leave the bathroom.
If I slide the turtle under the door, **then** she will leave the bathroom.
If I slide the frog under the door, **then** she will leave the bathroom.
If I slide the real snake under the door, **then** she will leave the bathroom.

Collect data:
You test each hypothesis separately and combine all the data on a chart.

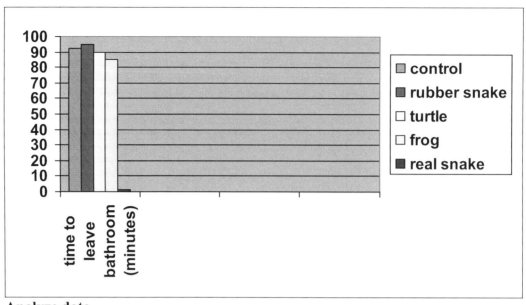

Analyze data

This part answers the hypothesis. The graph tells you that the control (when you did nothing but wait patiently) was 92 minutes. The rubber snake actually was worse at 95 minutes because she spent 3 minutes laughing and calling you a dweeb. The Turtle would not fit under the door, but did stick his little head under where she could see it, and although the time was reduced to 85 minutes, it was not much of a difference and the turtle probably did not cause the change. The frog showed promise but kept hopping the wrong way back to you. The real snake on the other hand caused immediate results. She shot out like a rocket. *Your last hypothesis was correct.* Congratulations on your scientific achievement.

Conclusion:

It is time to answer the original problem, and you know the answer! To get you sister out of the bathroom the fastest, slide a real snake under the door. You can then report this to your friends on the Internet!

Variables

People use the word variable as if there is only one kind, and then they confuse us when they talk about three different kinds. What is going on? Well a variable is what you *change* in an experiment, it could also be the things you *could change*, but do not. So to keep things as simple as I can, we will say everything you could *potentially* change in an experiment is a variable. The one you actually do change I will give a better name, the *Independent variable*.

Independent variable

This is what you (the scientist) actually choose to change in an experiment. It is the decision you made. The one thing you want to test, the first part of your hypothesis.

When you slid the snake under the door, you ***chose*** to do it. That was the *independent variable*. The frog, turtle and rubber snake were also independent variables because you **chose** to try them to see what happened. You can only have one independent variable at a time so you can't slide the snake and the frog under the door at the same time because you would not know which one caused her to leave the bathroom. The results or what happened is another kind of variable that you did not choose – the *dependent variable*.

Control variables (or Control)

These are the things you could change in an experiment if you wanted to, but decide not to change. Think of these as "normal" things. You might be doing an experiment on how to make bigger plants. You could change the amount of water but you don't want to, because you want to compare different amounts of fertilizer. All the things you keep the same (water, light, temperature, etc.) are *control variables*.

Dependent variable

This is what happens *because* of the independent variable. Notice you did not do this, you did not even know it would happen. It just happened, not because of you, but because of the *independent variable* you chose to do. *It is just what happened.*

When used together the independent variable is like a hypothesis but it is not making a prediction, it is *describing* the experiment. It is a cause and effect type of thing.

The independent variable *caused* the dependent variable.

A snake under the door *caused* my sister to leave the bathroom

A rubber snake under the door *caused* my sister to laugh and call me a dweeb.

Driving while texting *caused* the car to wreck.

Studying for my test *caused* me to get a good grade.

Eating the brown stuff *caused* me to throw up.

A good way to not get these switched is to remember:

I the scientists change the "**I**" for **I**ndependent variable.

A day in the life of Earwig Hickson III

One day I was going out for a walk and saw a small fluffy animal, how cute it looked, how friendly. I began to use the scientific method.

Problem (question)

I wonder if it would make a good pet.

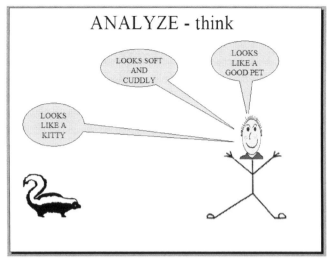

Research:

It looks fluffy like my stuffed toy animals. It looks soft like my blanket. It is colored black and white like nothing I have ever seen before. It kind of looks likes a cat and they make good pets. It could be some kind of cat. I wish I had a field guide to look it up. My parents would enjoy a new pet.

Hypothesis

If I pick him up **then** he will love me.

Collect data

I picked up the little animal and it sprayed me with really STINKY oil.

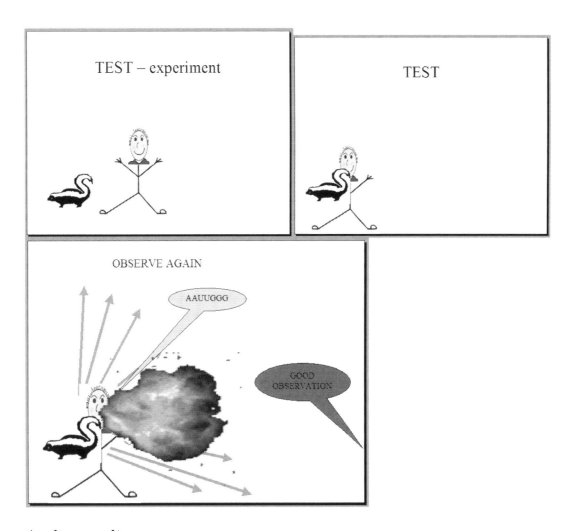

Analyze results

The experiment was a total fail. I am major STINKY from the thing spraying me. When I picked him up, he did not love me.

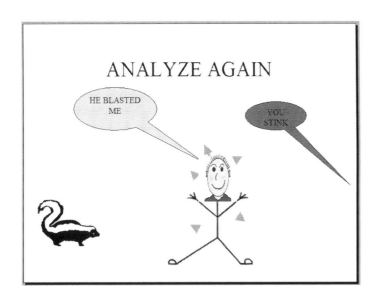

Conclusion:
It would not make a good pet.

When I got home I had a lot to ponder while I sat on the living room sofa watching TV. What was my mistake? I think I should have done a lot more research. A field guide might be a good investment. The Internet would have been nice. Asking someone if they ever saw one of those things could have helped. My hypothesis could have been better too. Maybe instead of picking him up I should have just tried to pet him. My experiment definitely could have been improved. I could have worn a raincoat and face shield. Maybe I should have made my friend Otis do the experiment while I observed. When my parents got home I was in for a surprise, I had another problem to solve, more research to do, another hypothesis to test, another experiment to do. All because I STANK BAD and so did the sofa and every room of the house I ran through. So did the tent they banished me to, until the smell went away. I liked the peace and quiet of the tent though; it helped me concentrate as I plotted my next attempt to make a pet out of the black and white, furry (and rather smelly) cute animal.

Useful stuff to know

Earwig could have neutralized the skunk oil odor very easily, rather than spend that winter living in a tent in the back yard with the following formula:
1-quart hydrogen peroxide
1-cup baking soda
1-tablespoon dish soap (Dawn is good)
Lather up and rinse. Repeat as necessary. This will actually oxidize the skunk oil into a new molecule that does not smell bad. Just in case.

Review of terms – Quizlet
https://Quizlet.com/124228565/chapter-2-volume-1-the-scientific-method-flash-cards/?new

Fun things to google

Myth busters - did man land on the moon?
Moon landings
Gemini project
Apollo project
Any Myth Busters episode
Thomas Edison

Links

Controlled experiment - https://www.youtube.com/watch?v=Hj3WkGLs7d0

Monty Python and the Holy Grail scientific method:
https://www.youtube.com/watch?v=GUQUqV0_PTc

Rats playing basketball: https://www.youtube.com/watch?v=jAQSEO25fa4

An experiment on gravity hill – where things roll uphill! – Every area has a *gravity hill*. There are myths that go with this, but has it ever been tested. Yes. I tested it!
https://www.youtube.com/watch?v=bsMZkm66uSI

Phet – color vision – every change you make is the independent variable, the resulting color is the dependent variable.
https://phet.colorado.edu/en/simulation/color-vision

Oast stick – This is a magic trick a magician might do, so you do not get to see how it works, but can you figure it out?
https://www.youtube.com/watch?v=5HyWbJpL2eQ

Defective paper – just an illusion but can you figure it out?

https://www.youtube.com/watch?v=sWeag7-wh3Y

Controlled experiment – Our two friends talking about how a controlled experiment works https://www.youtube.com/watch?v=Hj3WkGLs7d0

Here is a fun one – problem: How do I keep a squirrel out of the bird feeder? https://www.youtube.com/watch?v=gNB4WAPMwj0

A bird uses the scientific method to use bread to catch a fish: https://www.youtube.com/watch?v=uBuPiC3ArL8

Myth Busters use the scientific method to prove there are man made objects on the moon: https://www.youtube.com/watch?v=VmVxSFnjYCA

Big Bang theory repeats the moon experiment: https://www.youtube.com/watch?v=_v52LFgUq-8

Chapter 3
Physical Science

Movie of this chapter's power point lesson:
https://www.youtube.com/watch?v=39BKokdN9V4

The wonders of science

When you understand something, you can control it. When you understand the causes of a problem, you can solve it. Is air pollution a problem? Find the source and solve the problem. Is disease the issue? Find the cause and solve the problem. Understanding the world around you allows you to make it better. A good place to start is with Physical Science.

Physical Science

So what are we studying this year, Science? Well there are a lot of kinds of science. **Biology** studies living things. **Geology** studies the Earth. **Astronomy** studies the stars. Since we will be studying physical science, what is that? *Physical scientists study matter and energy*. It is a big field; there are a lot of topics. Many of the other kinds of science fall into or at least over lap with physical science. We are the *kings* of science.

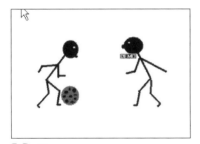 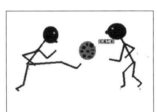 

Matter

We study matter, and matter is *everything*. Almost everything we know of in the universe is made of matter. Energy is not made of matter. Heat is not. Gravity is not. Light is not. Sound is not. The vacuum of outer space is not matter. But if you can put it into a jar, screw the lid on, and check back in a week, if it is still there, it is matter.

Matter is stuff. It has **mass** (or weight if on earth) if you get hit by a big chunk of it you would feel it. You can taste matter, smell matter, feel matter, see matter, but you cannot hear matter, but you can hear the sound matter makes. Everything is matter. You are matter, so you matter.

The study of matter is called **Chemistry**. Chemistry is when scientists try and find out what matter is made of. It is not easy to do this because the smallest pieces of matter are much too small to see, even with a microscope. At its basic level matter is made of atoms. **Atoms** are the building blocks of all things. There are only 92 kinds of atoms found naturally on earth, and there are not that many more kinds found in the whole universe. Atoms are found on the **periodic table of the elements** because each kind of atom is an element. Some of these you have heard of, Oxygen, Hydrogen, Gold, Silver, Iron, Copper, Helium, Carbon and many more common types. You can have a big chunk of the element Gold but the smallest piece you can ever have is an atom of Gold.

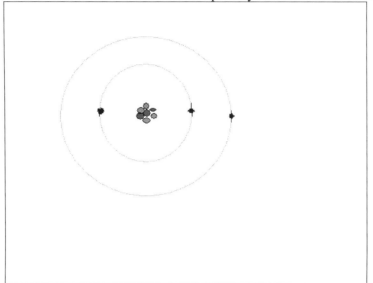

Atoms have a tendency to attach to each other and form bigger pieces called molecules (more on that later). Most molecules are not on the periodic table and those whose names appear there, are made of only one *kind* of atom, like Oxygen or Gold. Most molecules you have heard of are compounds and they are never on the periodic table. Things like water, salt, and rock candy are molecules and are not on the periodic table. The only time you will find water on the periodic table is if you spill it there. There are billions of kinds of molecules.

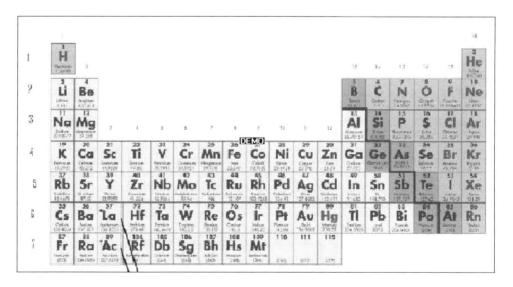

The other thing about matter is that it has **volume**; this means it takes up space, some of it is very small, like an atom, but it takes up some space all the same.

So *matter has mass and takes up space*, but often times you do not notice it. Did you know that air is **full** of matter? We just can't see it. So our jar may look empty, but actually it is full of air molecules. This is why a big gust of wind can knock you down and why hurricanes and tornadoes cause so much damage. Air molecules have mass and can cause really big boo boos.

A neat thing about matter is that it can change into something that looks completely different. You would not recognize it, it is almost like magic. What is odd though, is that no matter how different it looks, the *original atoms are still there*! You can't just get rid of matter, but you can *change* it. Want to hear something wild? If you go camping and burn a 50-pound log in the campfire, it *does not go away*. So where does it go? The log may be gone but the atoms it was made of are still somewhere, you just do not notice them. The atoms of the log turned into atoms of smoke and ash. They are made of the *same atoms* but are attached in different molecules. The amazing part to me is that that 50 pound log actually turns into 50 pounds of smoke and ash. The smoke just floats away and the ash hardly weighs anything. When you eat a pound of m&m's, they change into something else too, which you usually just flush away. The atoms are there, but it is not m&m's anymore. Trust me on that.

So what do we know about matter? It is everything that *has mass and volume*. Chemists study it. It can change but *never goes away*, or be created for that matter. *There is no such thing as magic, only science.* Scientists like to call this idea the **Law of Conservation of Matter**, it means that matter cannot be created or destroyed, only changed from one form to another. The universe recycles matter; it is not making it anymore.

The recycling of an atom

Let's think about an atom of carbon (one that is in all living things). This particular atom is found in an apple. We will call him **Adam**. A deer ate Adam one day and the deer digested Adam's apple into tiny individual pieces by the deer's stomach and intestines. Once in the deer, the apple was broken down into small molecules, some of these were used to make deer parts (meat, hair, etc) but not Adam; he was in a sugar molecule. Since Adam was in a sugar molecule (glucose if you remember that from life science) he was used for energy to make the deer move and stay warm. Through the process of **respiration** (the opposite of **photosynthesis**) the bonds in the sugar molecule were broken (producing the energy) and our carbon friend Adam was separated and all alone. Immediately two Oxygen atoms (from air the deer was breathing) grabbed him and attached to him with chemical bonds. This formed a new molecule called Carbon Dioxide (CO_2), just like all animals (and us) the carbon dioxide molecule went into the blood and into the lungs and the deer breathed it out. Adam was free! Into the atmosphere he went; drifting with the wind, part of the carbon dioxide molecule. He was having a fine time floating around until a corn plant absorbed the carbon dioxide he was in, and trapped him. Adam was used to make sugar again (remember photosynthesis); he ended up in a corncob kernel. A deer ate him and the whole process started again. Adam did take a short break as a little pile of deer poop, but was eventually eaten by the worms and went back into the air as CO_2 again. This is called the carbon cycle. The idea is that *no new carbon molecules are created and none are destroyed, they are recycled.*

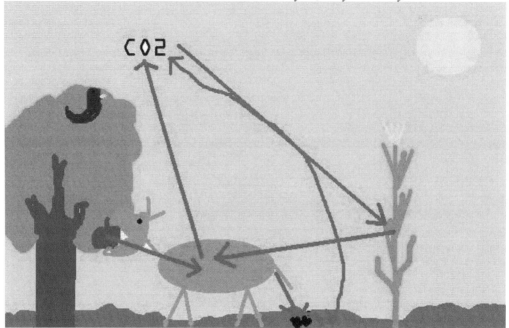

Energy

If matter is everything what is left to be energy? Well there are a lot of things you can put in a jar that are *not there* a week later, this is energy. Try shining a flashlight into a jar for an hour and quickly put the lid on. Can you open up the jar a week later and let the light out? Can you flash it in your friend's eyes? The light is gone because there was no matter in it. Light has no mass, it is energy. You can try and scream really loud into the same jar but a week later when you open it up to let it out, no scream. Gravity will not stay in a jar either. Only matter will stay in a jar. The study of energy is called *Physics*.

Even though energy is not matter it affects matter in such a way that we notice it, at least sometimes. We see light because it causes the nerves in our eyes to send a message to the brain. We notice sound because it vibrates air molecules and our eardrums. We notice heat because it makes our molecules go faster and our nerves (which can detect this) send a message to the brain that says HOT. We notice gravity when we fall down the steps. *Energy changes matter*. Sometimes it makes it change its motion, like when you fell down the stairs. Sometimes it changes it into something different like the log turning into smoke molecules. Sometimes it changes ice into water. Sometimes it does amazing things like when electricity makes the TV show a movie. Energy causes *all change* and if there was no energy on earth (including the sun) nothing would be happening, no weather, and no motion, no anything. Earth would be a dead chunk of rock. Nearly all the energy on earth comes from the sun. Some comes from the gravity of the planet or radioactive materials, but what you are used to, all used to be sunlight.

All sunlight? What about the energy you get from eating junk food, you know the stuff you are not supposed to eat but everyone does anyway. What is the energy in a piece of m&m candy? Well right now it is chemical energy trapped in the food molecules (by chemical bonds) but it was not always that. About a year or so ago, that energy was sunlight. The sunlight fell on a cocoa tree somewhere in Africa, which trapped the light and turned it into sugar through the process of **photosynthesis**. The sugar was stored in a cocoa bean (the bonds between the atoms) and was eventually picked and processed into m&m's candy. Do you know what that means? When you eat candy you are actually eating old *golden drops of sunshine*! How can eating sunshine be bad? Ask your health teacher that one.

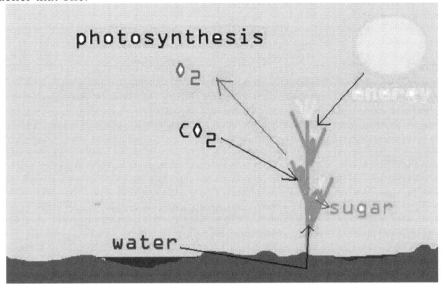

Electricity is needed for many things in your house to work, like the lights. Much electricity is made by burning coal and converting the heat into electricity, by turning a generator. The coal was at one time a bunch of ferns and other plants, 100 million years ago! Guess where the ferns got the energy? Yep, the sun, so when you turn on a light you are using the energy of 100 million year old sunlight. Oil and gas are the same way. In fact everything is. Not so fast you might say, what about hail falling from the sky? That definitely has energy because it hurts, where did that energy come from? You know the answer already; you just do not realize it. Hail is frozen water, water goes up into clouds by evaporation and guess what make the water evaporate, Sunlight!

You may have noticed that energy is in all kinds of forms but almost all came from sunlight. Is there something else strange about energy? Oh yea, just like matter it *never goes away* either, but it can change from one kind to another, you just do not notice it. When you shine a light into a jar where does the light go? It changes into something new; some of it turns into heat and makes the jar hot. Some of it shines through and is absorbed by the walls of your house into heat again. The heat then radiates through the walls to the outside, where it warms the air. This is one-reason cities are warmer than the surrounding farmland, and why a burning light bulb feels warm. The energy we use everyday changes into other forms, like heat. We just do not notice this extra heat.

There is a law to describe this too, the **Law of Conservation of Energy**, it means *energy cannot be created or destroyed, it can only change form*. The universe is not making energy anymore, just moving it from one place to another.

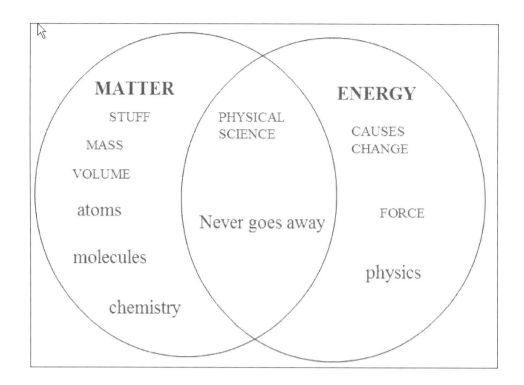

Force

One type of energy you do notice can be applied as a force. A force causes matter to **accelerate**, which means to change its velocity, which means to go faster, slower or turn. It is described as a *push or pull on matter*. And like all energy it *causes change*.

A day in the life of Earwig Hickson III

One hundred and seventy years ago, my great great great grandfather, Billy Hickson was conducting an experiment. He wanted to see where molecules go. He picked a local water molecule he had in a glass with some other water molecules. While it was sitting on his desk there was a knock at the door and in came his friend Abe Lincoln, who was returning a penny he owed him from earlier in the day. Abe had just walked three miles to return the money and saw the science experiment on the table and since he was thirsty, he drank it. My, was my great great great great grandfather mad! With the experiment ruined all he could do was start again, or should he? Where exactly was that water molecules, where would it go? He decided to keep an eye on it. Well the glass of water was too much for Abe and soon he had to visit the little outhouse out back and guess what? Out came the very same water molecule! The experiment was saved! Since he was still upset with Abe for drinking the experiment he sent him home. Abe was so upset he quit his job and moved away. My great great great great grandfather never saw him again. He heard that Abe got into the politics game after that and moved to Washington. He did not know what ever became of him.

Anyway back to the water. It slowly moved in the dirt from the outhouse to the garden where it was sucked up by a tree root, went up into the tree by the process of **transpiration** and out the leaves, and up into the sky. It got together with some other water molecules and soon was part of a cloud. The wind blew the cloud to the west where the water molecule fell out with a bunch of others in the form of a raindrop. It landed on the side of a mountain and got mixed up the more water and rolled down in a river. A fish sucked it in its mouth and out its gills. Then a deer drank it and squirted it out on the ground again. It evaporated back up into the sky where it fell as snow on some kid's tongue. He swallowed it and it ended up in an outhouse again. For the next 50 years it bounced around in something called the *water cycle*, constant evaporation, condensation, precipitation, and then evaporation again. Over and over until great great great great great grandfather lost track of it when it flowed into the ocean and sank. Our family was kind of glad when the experiment ended, since my great great great great great grandfather kind of wasted 50 years of his life following a water molecule, rather than get a real job, but that is how some scientists are.

I tell this story now because a funny thing happened today. I found the water molecule; the very same one Abe Lincoln had drunk! It was in my kool-aid! It had evaporated out of the ocean and fell as a raindrop and landed at the city reservoir. When I got some water to make my drink, it was there. I drank it before I could appreciate the implications that I just drank part of Abe Lincoln's urine.

It kind of makes you wonder who or what drank your water doesn't it. Are there molecules that passed through a dinosaur? George Washington? Me? There is no "new" water just the old recycled stuff that has been here all along.

Note

In reality, although the water cycle is the recycling of water, sometimes water molecules are split into hydrogen and oxygen. This happens in plants through photosynthesis when the atoms are used to make glucose. Animals actually make water when they use glucose for energy through the process of respiration (some animals make their own water and never have to drink- google kangaroo rat). You could split water yourself by electrolysis (google that too). The idea of the story is still valid though, it is just that instead of the water molecules being unchanged it would be the atoms of hydrogen and oxygen, it was made of.

Useful things do know

Even though you drink the same water that someone up river flushed away, we treat dirty water in a water purification plant and what comes to your house has all the bad stuff filtered out. If you have a well, the ground filters it before you drink it. So don't worry that you are drinking used water just be glad science had figured out how to clean it and make it healthy.

Review of terms – Quizlet:
https://Quizlet.com/124252137/chapter-3-volume-one-physical-science-flash-cards/

Cool things to Google

Electrolysis of water
Water cycle
Bill Nye energy and matter
Drinking water treatment
Kangaroo rats do not drink water in the desert
How do solar panels work?

Links

Conservation of matter with Napoleon -
https://www.youtube.com/watch?v=4PJOq30dwvo

Conservation of energy with Lincoln -
https://www.youtube.com/watch?v=qQP3FeDXhr8

An example of energy (light) making matter (radiometer) move. It looks like magic but it is science: https://www.youtube.com/watch?v=CV_aGapVyaM

A cool demonstration where the energy from super hot steam is us used to light a match: https://www.youtube.com/watch?v=73k9DuAS9v4

This is a famous demonstration called the Gummy Bear experiment. This is kind of what your body does when you eat a gummy bear, it uses oxygen to burn it, only slower. https://www.youtube.com/watch?v=Ca4kY5gGI4c

Chapter 4
Lab safety

Movie of this power point lesson:
https://www.youtube.com/watch?v=8RuUXBzNeyc

The wonders of science

Look before you leap. Know what you are doing before you do it. Do not do anything stupid. An important rule in science is to live another day to do more science. Not all scientists did this. They did not last long.

How did people figure out what foods were edible and which were poisonous? I do not think people played with their lives when it came to eating a leaf or a fish. They followed common sense, and we now call this lab safety. I am sure primitive people had safety in mind whenever they tried something new. Imagine how the first hunters to kill a Wooly Mammoth went about it, I doubt they jumped on it and wrestled it to the ground. How did they domesticate wolves into early dogs? They were careful, they wanted to live. Evidently they did things right.

Lab safety

There is nothing more harmless than a grizzly bear, except when they are chewing on you. That is how it is with chemicals. Chemicals are harmless as long as you use them correctly. The chemical, *water*, kills more people than any other chemical in the world because people are not using it right; they try and breathe under it. A Rattlesnake is as harmless as anything, except when it is biting you. So do not let it bite you.

All chemicals, like all rattlesnakes are harmless when used correctly and dangerous when used incorrectly. So use them correctly! In science we use safety methods to assure no one gets hurt. The word "chemical" scares some people, these people are idiots. All chemicals are harmless when *proper safety protocol* is used.

In a science lab there are certain expectations that *must be followed*, these are to insure safety. Common sense is really what we are talking about. Anyone can be an immature fool; the trick in science lab is to be mature. No one wants to sit with a dangerous person. No one wants lab privileges to be taken away because of one person, but this can happen. One immature person is a *danger to everyone*.

Common safety rules you may be required to do are as follows.

There is **no horseplay** in the lab. Keep your hands to themselves. This is not the time to show how funny you can be. Make sure you know the difference between humor and immaturity. Pipettes (eye droppers) are not for squirting someone. You may think you are the only one who ever thought of that, but in reality you are not. You are being stupid. Fun things happen in a lab without you having to add your own "humor". The lab is not the place to be funny. Remember maturity rules. No one will want to be with you if you act immaturely, so don't.

Imagine the bad things that can happen if you get bumped into from behind into what ever it is you are experimenting with; it might be a burner or a nasty chemical. You want everyone in the room to behave.

Safety glasses are a big deal in middle school science. You only get 2 eyes in your life, so do not waste them. I always recommend splash goggles, they protect from the front and the sides. The accident usually comes from the sides (where the idiot sits) not the front. Safety glasses are the prime defense against these idiots. One of the few times I really yell in class is when someone takes off their safety goggles. They are the first line of defense and work very well. They also make you look smart. If you need to remove them for any reason, wash your hands and leave the room to fix them.

Sometimes you may wish to wear an apron. These are usually not necessary, but if you have extra nice clothing on, you may wish to use one.

Never taste anything in the lab. For one thing, do you know what has been in these beakers in the past? I know I do not. Maybe they were for collecting urine, maybe they held nasty bacteria, that can give you industrial strength diarrhea, I do not know and neither do you. So do not taste anything in the lab. Do not even eat in a lab. It is a bad idea.

Smell things by wafting, this is when you wave your hand over a chemical to mix air with it before you smell it. It is much safer than smelling the concentrated chemical. I once had a superintendent come in half way through a lesson and sniff a beaker of pure ammonia. He pretty much burned his nasal cavity. I got in major trouble because he considered that a dangerous chemical to use. None of my students sniffed the ammonia. They waffled. What an idiot.

You do this naturally when ever you smell a strong odor, it is a survival instinct.

Fire is a reality, and if you do not know about fire, it can burn things, and you are flammable, especially, your hair. For some mysterious reason, teenagers think that putting moose and other forms of hair gel on their hair is important. All this stuff is made from *crude oil*, and crude oil is very flammable, that is why we make gasoline from it. That makes hair a very good way to start a fire, good if in the wilderness, but bad if in the lab. So a smart person would **tie back long hair** to avoid a "fireworks effect". The burner should also be in the middle of the table where you can control it, and no one else can lean back into it!

Always remember and never forget, hot glassware looks just like cold glassware. The difference is that if you pick up a cold test tube, no problem, but if you pick up a hot test tube, big problem, because your reflexes are going to make you throw it. This is not good.

Always listen to the teachers, they will be giving instructions that include the proper way to treat chemicals and do a lab safely. They know what the dangers are so listen!

Stay at your lab station, there should be **no unnecessary walking around**. There is enough necessary walking around. If someone across the room had a cool experimental result, do not run over to see it, you will get the same cool result at your station. Most accidents happen at the beginning of a lab and during clean up when people must walk around the room. Be careful at *these times and keep your safety glasses on.*

Safety equipment in the lab

Each lab should have an **eyewash station**; this is for when the safety goggles fail for some reason. It includes pushing your face over a squirting water column of torture, to wash out anything in your eye. It is not fun and usually lasts about 15 minutes. To avoid this *keep your goggles on*!

There should be a fire blanket in the room. In the unfortunate situation that someone catches on fire (and clothing is flammable too) this is the final line of defense. You simply cover the person with the blanket so the lack of oxygen makes the fire go out. Since most people, when on fire, have a tendency to panic and just pick a direction and run, you may have to catch them, and cover them in the blanket, it works but if you follow good safety rules should not be needed.

Each room should have a fire extinguisher. Again this is usually not necessary if proper lab behavior is followed but just in case. Make sure you know where the fire extinguisher is and how to use it. During an emergency is not the time to learn how to use one. In a nutshell – pull the pin and squeeze the lever.

In summary, **never remove your safety goggles**, **no horseplay**, **stay at your lab station**, **no unnecessary walking around**, do not squirt people with a pipette, in other words use common sense.

A final note: doing labs is a **privilege**, not a right. Labs are fun but one person can ruin it for everyone, if they are not safe. The unsafe person always *causes* the accident; the innocent bystander is the one, who gets hurt,

Remember
In other classes you can stub your toe, drop a book on your food, get you hair stuck in a pencil sharpener, or staple your fingers together. But in science you can get a burn that will not stop burning, catch your hair on fire, become blind. Behave in science lab.

A day in the life of Earwig Hickson III

Today I got stuck with the worst lab partner in the world, Parvis Nitwitrod. He is the worst kid in the school and ruined our experiment. The first thing he did was push my head toward the Bunsen burner as joke, I could see the flames against the lenses of my safety goggles, and I was glad I had them on. Then he screamed BOO, when I was trying to slowly add a chemical to the solution we were making, I spilled it all over the place, Parvis laughed. Later when we were supposed to be writing down data he went to the lab station across the room so see what they were doing, he tripped over a back pack and knocked a person into an odd purple liquid that splashed all over their nice Philadelphia Eagles jersey. That boy was upset and started yelling at Parvis. Things would have got out of hand except that Jumbo Wilson, the star football player, threatened to put Parvis through the wall if he did not get back to his station, I was rooting for Jumbo. While Parvis was gone I did get a lot of work done.

Parvis then proceeded to take his goggles off so he could scratch his eye; there was some nasty chemical on his fingers, which he rubbed into his eye. It must have burned a lot because he screamed like a little girl. He spent 15 peaceful minutes at the eyewash station. I enjoyed that.

When we made a cool liquid that suddenly turned purple, he drank some! Nothing happened at first but I heard later that he spent a long time in the bathroom producing purple diarrhea, funny stuff for me.

The day ended with Parvis taking his safety glasses off while heating the purple liquid. He was very surprised when POOF, he no longer had any eyebrows.

The poor teacher tried to keep an eye on Parvis but every time he turned his back, Parvis was back at it. I think our lab privileges are going to go away after today.

Review of terms:
https://Quizlet.com/124252991/chapter-4-volume-one-lab-safety-flash-cards/?new

Fun things to Google
Lab accidents
Michael Jackson Pepsi commercial
Lab safety

Links
One of my experiments gone wrong
https://www.youtube.com/watch?v=sSRUnUj7TOU

Using an eyewash station: https://www.youtube.com/watch?v=obO8Sq9xSMQ

Chapter 5
The Controlled experiment

Movie of this chapter's power point lesson:
https://www.youtube.com/watch?v=zkzekWwZDNw

The wonders of science

Long long ago some caveman figured out how to start a fire with a fire drill. Think how hard that was. What kind of wood is best? What kind of tinder is best? Some ancestor of ours compared different materials and techniques until the rather difficult skill was invented. This is how a controlled experiment gives us knowledge.

How a fire drill works, https://en.wikipedia.org/wiki/Bow_drill
My fire bow: https://www.youtube.com/watch?v=APCgA9uxMGI

The Controlled experiment

The controlled experiment is the corner stone of science; it is the only way to learn new stuff. A controlled experiment has **ONE independent** variable, which is compared to another experimental set up that is "*normal*" or has *no independent variable*... Everything else is the same. These things that are kept the same are called controls or *control variables*. Only *one* thing can be changed in an experiment, otherwise you would not know what caused the results. The controlled experiment removes the possibility of a coincidence.

At this point I should review the kinds of variables. The one thing you (the scientist) can change is called the **independent variable**. What happens in the experiment is called the **dependent variable**. The independent variable *causes* the dependent variable, and you can only have *ONE* independent variable. Sometimes you see the word variable by itself, this means the independent variable, and this confuses people.

There are a ton of things in an experiment that are not changed. These are called **control variables** (because you could have changed them) or just controls. It is important to keep everything the same in an experiment, *except the independent variable*.

The other very important thing about an experiment is that other scientists can copy what you did and get the exact same results. Scientists actually try and prove other scientists wrong and if they do, the other guy is toast. No one will ever trust him again. But if you prove him right, it is called verification, and it makes him look good. In other words, your job is to support another scientist or destroy him. This is why science can be trusted, we verify everything.

There once was a scientist from South Korea who was an expert on cloning. He achieved amazing results. He cloned a dog. He was ahead of everyone. But as other scientists tried to repeat his experiments they did not work. It took 5 years but eventually we found that he was faking his results, he was a fraud. He lost his job and all credibility. He now works for some organization searching for flying saucers. Science does not tolerate cheaters! And we eventually always catch them. One of the cornerstones of science is to report your results honestly. No one likes a cheater.

A typical controlled experiment

The simplest example I can think of is the old fertilizer experiment. Let us say we want to know if a particular fertilizer for plant growth really works. The first thing we would do is get a lot of plants because more is better. A thousand corn plants will give us much better results than only two. We keep everything exactly the same except for ONE thing, the *independent variable*. The amount of light, water, number of seeds, everything is exactly the same. If you wish to sing to your corn, sing to all of them, do not play favorites. The one and only thing that is changed is the fertilizer (the independent variable). You can compare your fertilizer to other kinds of fertilizer or to no fertilizer that is your choice. The independent variable is what ever YOU want to test.

So you plant 1000 corn plants in the same area, from the same seed packets and give them the same amount of water. You keep everything exactly the same; except you give half the plants the new fertilizer and you give none (or another brand) the other half. You now know the only difference between the two groups is the fertilizer. This means that any changes can be caused only by the fertilizer and nothing else.

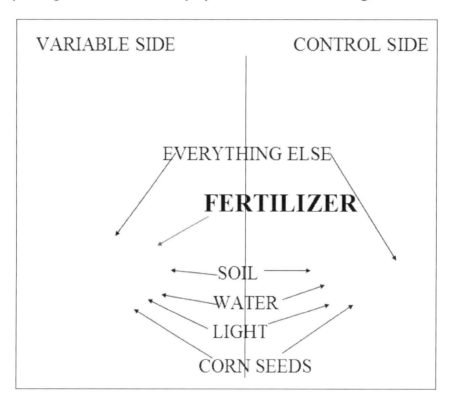

Your hypothesis was probably something like, "*IF I give some plants the new fertilizer THEN they will grow taller*". This is a great hypothesis since it can be tested. Your results will tell you if your hypothesis was correct. If the majority of plants got bigger – success! If they did not FAIL! Even failed experiments are successful, because you learned something.

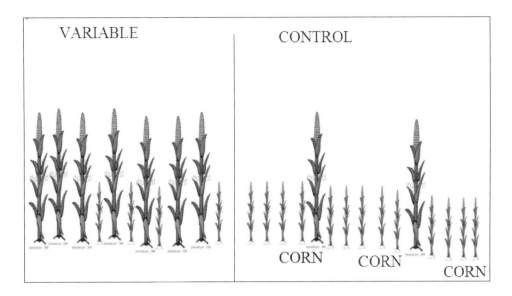

In our experiment the experimental group (the one with the new fertilizer) grew bigger. This implies that the fertilizer works. We learned something and we report it exactly as we found it.

The lab report

Unfortunately we have to report our experimental results. This is the part people hate, but it is necessary. This is our report of the previous experiment. Notice it is not long, it is to the point.

Question or problem: Does "Bob's super fertilizer grow bigger plants?

Research: A quick google search found that some people liked Bob's super fertilizer but some did not. Not much help.

Hypothesis: *If* I use Bob's super fertilizer *then* my plants will grow bigger. I will keep everything else exactly the same.

Collect data: The average size of the control group (no fertilizer) was 75 cm after 3 weeks. The average size of the variable group (with Bob's super fertilizer) was 125 cm after 3 weeks.

Analysis: the experimental results seem to show that using Bob's super fertilizer resulted in larger plant growth, 50 cm more in 3 weeks. Not bad.

Conclusion: Bob's super fertilizer does grow bigger plants.

A day in the life of Earwig Hickson III

Today I wanted to find out if doing homework actually helped my test scores. I do not like doing homework, but for some reason my teacher thinks it is important. I wanted to prove that homework is a waste of time and I should not have to do it. My hypothesis was that "IF I do *not* do my homework THEN my grade will be just as good as normal".

My control variable was that I did my homework for the first test. The result of the test was a 94%. Since this was my control, it is what my experiment was compared to.

My next test I did not do my homework (for scientific reasons only). The results were not good. I kept everything else exactly the same, I paid attention in class, I took my notes, I even studied my notes but my test result was only an 80%.

The difference between doing my homework and not doing my homework was 14%, the difference between not being grounded and getting grounded.

Maybe doing homework is more important than I thought. My teacher does say doing homework is important, maybe he is right!

I learned something, I should do my homework! At least when I was grounded I got to play with my new friend Tom the spider.

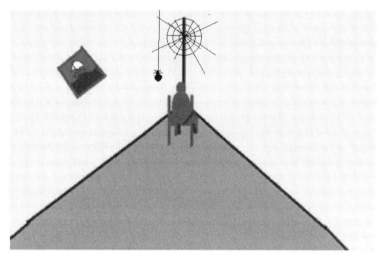

Review of terms – Quizlet:
https://Quizlet.com/124253531/chapter-5-volume-one-controlled-experiment-flash-cards/?new

Fun things to Google
How to do a science experiment
Controlled experiment
Mythbusters – moon hoax
Apollo moon missions
Gemini space missions
Did we land on the moon?

Double blind test
Controlled experiment

Links

Mythbuster's flag on the moon experiment
https://www.youtube.com/watch?v=JAC0NSFNEPQ

Mythbusters did we land on the moon
https://www.youtube.com/watch?v=VmVxSFnjYCA

Mythbusters interview about the moon-landing hoax
https://www.youtube.com/watch?v=jhuX-UaQiS4

A controlled experiment with Simon and Sara
https://www.youtube.com/watch?v=Hj3WkGLs7d0

Controlled experiment
https://www.youtube.com/watch?v=aLesk8fujH8

Scientific variables
https://www.youtube.com/watch?v=x2606GQmDqY

Unit 2
What is Matter?

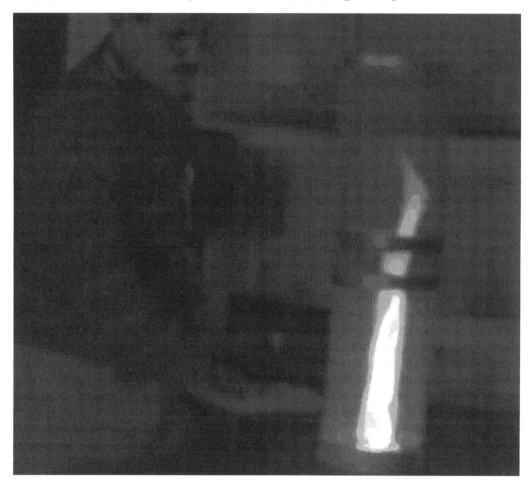

Chapter 6
Measuring Matter

Movie of the power point for this chapter:
https://www.youtube.com/watch?v=Qbwz8rvemn8

The wonders of science

When people started trading products a few thousand years ago they needed a way to calculate how much was owed. There was no money. Eventually gold and other precious metals were used to buy and sell products. Odd as it sounds gold is only valuable because everyone thinks it is. The problem quickly arose, how do you know how much gold you have. There needed to be a way to measure it. The answer was an accurate balance. People needed to measure things accurately and we still do.

What you should know

Matter is anything you can *put into a jar and take out later*. It is anything you can hold. It is everything you are familiar with. Matter is *stuff*. The scientist says that matter is anything that has **mass** and takes up **space** (has **volume**). If you can kick it, it is matter, if you can hold it, eat it, smell it, taste it, it is matter. If it is a solid, liquid or a gas it is matter. If you can put it in a jar and take it out later, it is matter. Matter is easy to measure too, you can measure its mass (with a **balance**), its volume (with a **graduated cylinder**), and its weight with a **scale**.

Things that are not matter are **energy**. Light is not matter because it has no mass or volume, or better yet, cannot be shined into a jar and let out later. Sound is not matter either. Matter can make sound but the sound itself is energy. You cannot scream into a jar, put the lid on quick and scare someone when you open it later. The sound is not there, it is not matter. Gravity is not matter either, it affects matter by making it fall but it is not matter by itself. You cannot put gravity in a jar, take it to the international space station, open it and give the astronauts gravity. It has no mass or volume, it is not matter. Energy causes matter to change but it is *not matter*. Energy causes matter to do a lot of stuff. It makes objects go faster, slower or to turn (this is *acceleration*). Light can make colors fade, sound can make things vibrate, and gravity can make things fall. Energy causes *changes in matter,* but it is not matter.

Mass

Since I just said that matter has mass perhaps I should tell you what mass is. Mass is actually *how much matter an object has*. This means how much *stuff*. More accurately mass is how many atoms are in something. Mass can be measured with **balance**. Do not confuse mass (the amount of matter in something) with weight. Weight is totally different. Weight is how hard gravity pulls on a mass. Many people incorrectly think mass and weight are the same thing and we will clear this up right now.

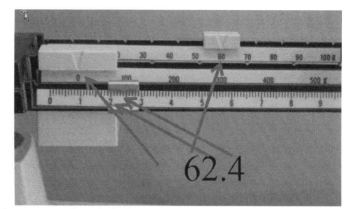

Triple beam balance   how to read a triple beam balance

Mass is how much *material*, **weight** is how hard *gravity* pulls that material. *Mass is measured with an instrument called a balance,* kind of like a see-saw (that compares an unknown mass to a known mass), *weight is measured with a scale* (which uses a spring or similar material). A scale is placed between the object and the earth and measures how much it gets squished. *Mass can never change* when moved to a different gravity field. *Weight changes with gravity.* If you went to the moon your weight would change (1/6 of your weight on earth) but your mass would stay the same because you have the same number of atoms as you did on earth. The *units for mass are grams* (about an m&m candy) or kilograms (about a brick); *weight is measured in Newtons* (a unit of force – about 1/5 of a pound). The American equivalent of a Newton is a pound, but the American equivalent of a kilogram is actually called a SLUG. We never use it, but that is what it is. Imagine if your mass was calculated in slugs at the doctor's office: You are 3.1 slugs! Not very nice sounding to me, my weight is bad enough! Do you want to know how many slugs you are go here: http://www.unitconversion.org/weight/kilograms-to-slugs-conversion.html

It is interesting to me that a Newton is about 1/5 of a pound; this is about the weight of an apple. It is said (probably not true) that Sir Isaac Newton was hit by a falling apple when he dreamed up his Law of Universal Gravitation. This would mean that *Newton was hit by a Newton.* (Groan)

Balance   *scale*

If you went to the moon you would have the same number of molecules in you, thus the same mass. But since the moon has less gravity it would pull on you less so your moon-weight would be 1/6 of earth-weight! If you were in deep space your weight would be zero, if you tried to weigh yourself on a bathroom scale, it would just float away, saying zero the whole time. You would be weightless, but not mass-less. You would still have the same number of molecules, thus the same mass. If you went to the bathroom in zero gravity, that would be weightless too, but since it is made of molecules it still has mass, so avoid it. Try this simulation, it will tell you your weight on other planets. Remember your mass would still be the same. http://solarviews.com/eng/edu/weight.htm

When I step on my cat's tail and it screams at me, it may not be my fault, it could be that the Earth is pulling too hard on me. My weight is caused by the gravity of the Earth.

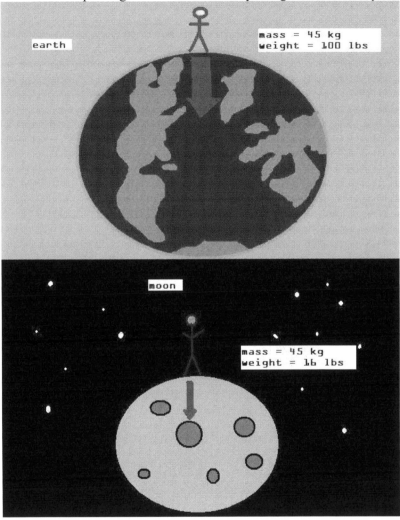

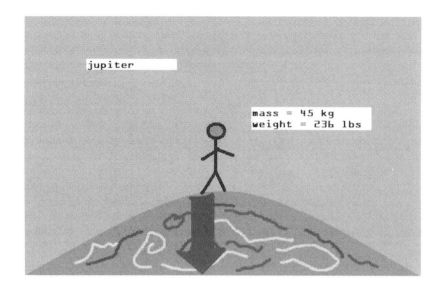

Earlier I said that mass (Kg) are measured with a balance and that weight (pounds) would be measured with a scale. This is true, but I have a scale that reads Kg and I have seen a balance that reads pounds. This is because these devices were made incorrectly and most people do not care. I call them Walmart specials.

Do you want to know how to use a balance? Go here. https://www.youtube.com/watch?v=FfuBO3-K8AQ or this cool simulation http://touchspin.touchspin.com/DisplayTBB.php

COMPARE AND CONTRAST

MASS	WEIGHT
MATTER	GRAVITY
BALANCE	SCALE
CANNOT CHANGE	CAN CHANGE
GRAMS	NEWTONS

Volume

Since matter has mass (stuff) and takes up space (volume) perhaps I should describe what volume is. **Volume** is nothing more than *how much space an object occupies*. How big it is. A soda bottle holds a volume of 2 liters (2000 milliliters), an eyedropper just a few *milliliters*. A room holds a volume of air in the hundreds of *liters*. *Size is volume*. In math you probably leaned to calculate the volume of a box using the formula:

Volume = length x height x width

In science volume is usually measured with a **graduated cylinder**. It is easy to measure the volume of liquids this way, but it can also be used to measure the volume of odd shaped solids. The method we use is *water displacement*. **Displacement** is how much something is *pushed away*. Measure a known volume of water in a graduated cylinder (say 30 ml), and put the object in with it, the amount the water that rises (displaced) is its volume. So if the water level rises to 35 ml, the volume of the object is 5 ml (35 – 30). Do you want to know how to use a graduated cylinder? Try this. https://www.youtube.com/watch?v=1od4Xgg2PZA

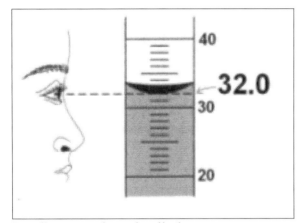
Reading a graduated cylinder

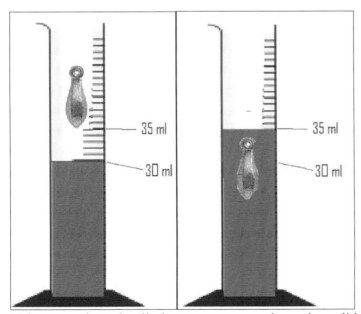
Using a graduated cylinder to measure an irregular solid

The standard unit for volume is the milliliter, which is the same as a cubic centimeter. Think of a ml as about the size of 1/5 of a teaspoon – so a teaspoon is about 5 ml.

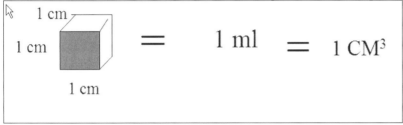

Review

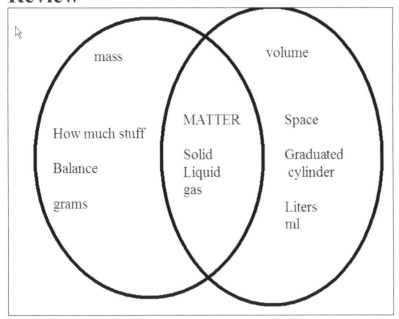

Density

The ratio between mass and volume is called **density**. Things with high density are heavy for their *size*, like lead or rocks; things that have a low density are light for their size, like feathers or foam. The density of water is 1 g/ml; this means 1 ml of water has a mass of exactly 1 gram. Things that sink in water have a density higher than 1 g/ml, things that float have a density smaller than 1 g/ml. Things with a *high density are compact, low density things are fluffy.*

A good way to think of this is with a balloon, When the balloon is empty it is dense, but when filled it is less dense because it is bigger but has about the same mass (the air does add a bit of mass but not much).

The molecules in *dense things are really squished together*, they are super smooshed and the molecules are very close together

A day in the life of Earwig Hickson III

My homework today was to use my science vocabulary words in the real world. My teacher said the more we use these words the better we will understand them. It did not go well.

I learned to be very careful with my new knowledge. Some of the things I should not have said were:

"My, what a tremendous *volume* you have today Mom!"

"I see you have significantly increased your *mass* the last couple of weeks, dad!"

"The *scale* broke from the force of *gravity* causing your *weight* to crush it, Mom."

"Sister, you have a big head without much in it, which makes you very low in *density*. I think your head could float in water!"

My Dad did help me review the term FORCE though, and how it makes little kids *accelerate*. He told me he was building my character, I think I have enough character already.

Fun things to Google
How to use a triple beam balance

How to read a graduated cylinder
How to find the density of a person
Phet simulations
Vomit comet KC-135 – how astronauts practice in zero gravity

Review of terms – Quizlet:
https://Quizlet.com/124254127/chapter-6-volume-one-measuring-matter-flash-cards/?new

Links

Your weight on other planets
http://solarviews.com/eng/edu/weight.htm

Converting Newtons to pounds – how much do you weigh.
http://www.convertunits.com/from/newtons/to/lbs

Triple beam balance simulation – touchspin – helps you learn to use a triple beam balance.
http://touchspin.touchspin.com/DisplayTBB.php

Since weight can change from one gravity field to another, what would your weight change to on other plants? http://solarviews.com/eng/edu/weight.htm

A story about Archimedes by Ted-ed. http://ed.ted.com/lessons/the-real-story-behind-archimedes-eureka-armand-d-angour

The crown of Syracuse, Archimedes great discovery about water displacement.
https://www.youtube.com/watch?v=wEvtahSn_ms

How heavy is air? A film from Ted-ed. http://ed.ted.com/lessons/how-heavy-is-air-dan-quinn

Why does ice float on water. A film from Ted-ed. http://ed.ted.com/lessons/why-does-ice-float-in-water-george-zaidan-and-charles-morton

Using density to separate two kinds of beads. Some have a density more than water and some less: https://www.youtube.com/watch?v=AS5xUhNkZhI

Chapter 7
Atoms

Movie of the power point for this chapter:
https://www.youtube.com/watch?v=_eIYOrEztqE

The wonders of science

It is an on going process. What is stuff made of? At one time people thought everything was made of a mixture of earth, water, air and fire. People dabbled in the idea that worthless stuff could be changed into gold, how far we have come. Now we are trying to find out what atoms are made of and what the parts of atoms are made of and what those parts are made of. We have much farther to go.

Alchemy, the beginning of chemistry, https://en.wikipedia.org/wiki/Alchemy

Atoms

You have heard of atoms but what the heck are they. You may know they are small, but how small. You may know they make up everything, but how do they do that? The word atom actually means, "that which cannot be split." A Greek chap named Democritus thought the idea of this up over 2500 years ago. He realized that if you chopped a piece of gold or something in half enough times, you would come to a point when you could not cut it again, no matter what tools you had. He thought there had to be an end to cutting the gold. There had to be a smallest possible piece of gold. He was correct; this smallest piece is called an atom.

Well mostly he was right. He was totally right if you cut the gold with physical methods (like chopping) or even chemical methods (using chemical reactions and energy). Using "non-nuclear" methods the smallest piece of matter you can get to is an atom. The names of all the atoms are found on the periodic table of the elements and some of them you have heard of: Gold, Oxygen, Hydrogen, Helium, and Aluminum. There are about 92 kinds of atoms found on earth, and some extras that have been made in the laboratory.

Interestingly, many things you have heard about are not types of atoms at all; they are made of those same atoms though, just put together into compounds. You will find none of these on the periodic table. They include things like water, air, wood, rocks, salt, birds, kitchen sinks, cows and even people. We will discuss these later, for now let's stick to atoms. https://www.youtube.com/watch?v=o-3I1JGW-Ck

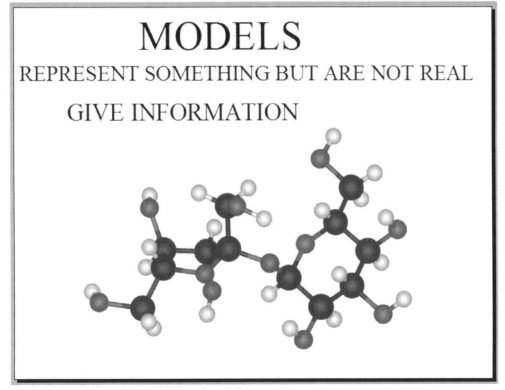

Periodic Table of the Elements

EACH OF THESE IS A KIND OF ATOM

Atoms are the basic units of matter

Think of the kinds of atoms on the periodic table as a kit. A kit that can be used to make everything in the universe! Using this kit you could use 1 atom of oxygen and 2 atoms of hydrogen and make water, which is the compound H_2O. You could also take a tree and cut it into all the building blocks of atoms, you would end up with a big pile of hydrogen, oxygen, and carbon, along with smaller piles of nitrogen, phosphate and dozens of other kinds of atoms.

MODELS
REPRESENT SOMETHING BUT ARE NOT REAL
GIVE INFORMATION

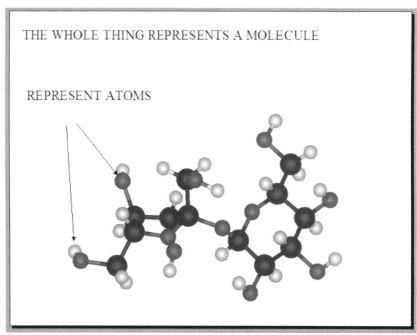

THE WHOLE THING REPRESENTS A MOLECULE

REPRESENT ATOMS

It is kind of like taking a Lego set and making a really huge castle. We would all agree it is a castle and not a Lego, but when we knock it down we would have piles of all the original Lego shapes and colors, but no more castle.

The universe is kind of simple after all! To make your own universe you would need a kit of 92 containers each filled with one of the 92 kinds of atoms. If you had your own "make your own universe kit", you could make anything you wanted. These kits are hard to find though, even on Amazon.

How small are atoms?

Really small, no I mean really, really small. Let me rephrase that atoms are really, really small. Way to small too see with a microscope. The average size of an atom is about 0.0000001 mm in diameter, what does that mean? Well if you had a piece of cardboard 1 mm thick it would be 10,000,000 atoms thick – 10 million! That is small. Some people like to compare the number of atoms in a grain of sand as being a lot more than all the grains of sand on a beach, a really big beach. The period at the end of this sentence probably contains about 5 million atoms of ink. If you want to see a great video that can explain this better than me Google "just how small is an atom TED (http://ed.ted.com/lessons/just-how-small-is-an-atom)", it will help you to understand. Can we see atoms? Not really but ….
https://www.youtube.com/watch?v=yqLlgIaz1L0

What are atoms made of?

Well you know now that atoms are really small and are the basic units of all matter. You can't cut anything smaller than one atom. Or can you? With our technology we can use a method called a nuclear reaction or nuclear change to actually see what atoms are made of. Scientists have developed experiments and amazing technology (like the Large Hadron Collider) that give clues about what an atom is made of. I must stress that we cannot see what is in the atoms, only measure them and take a very good guess at what is inside.

Some very smart scientists designed some very clever experiments that gave very good hints about what an atom is made of. J. J. Thomas discovered negatively charged particles he called electrons. It turns out electricity is moving electrons. Ernest Rutherford discovered a positive particle in the nucleus we call a proton. James Chadwick found a neutral particle in the nucleus we call the neutron. These were very clever people, and you probably do not appreciate them much but they are well worth your time on Google. Smart people deserve some Google time too! They improved your life in ways you can never really imagine.

So atoms are made of protons and neutrons in the nucleus (or center) surrounded by fast moving electrons on the outside locations called shells or energy levels. It is really quite simple. By the way a shell is not a physical object it is more of a location where electrons are found. Do you want a simple explanation of what atoms are made of? Try this. https://www.youtube.com/watch?v=sRPejoNktKE

Now if you could not afford the giant 92-part universe kit and you needed to save money, you could get the "budget make your own universe kit". This kit only contains 3 containers, one for protons, one for neutrons and one for electrons. From this kit you can make ALL the kinds of atoms, and then put them together as you please. The only difference, it turns out between the kinds of atoms are the *number of protons in the nucleus*. You heard me; the only difference between atoms is the *number of protons in the nucleus*. Atoms are actually very simple. Hydrogen has 1 proton, helium 2, lithium 3, and so on.

If you want to try and make your own atoms, this is a cool simulation from Phet, https://phet.colorado.edu/en/simulation/build-an-atom.

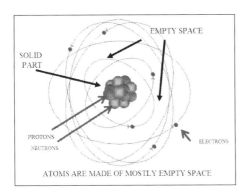

Now for something really weird

An atom has mass, but nearly all that mass comes from the nucleus, which is smaller than even the atom and we know how small that is. Since all that mass is located in such a small part of the whole atom, what is the rest of the atom made of? The answer is NOTHING. The volume of an atom is mostly *empty space*! There is nothing between the nucleus and the electrons and mostly nothing between atoms in an object. If you really want to impress your teacher or parents explain to them that their brain is made of mostly empty space. Your teacher will know you read this chapter and your parents …. Well good luck with them.

Let us magnify a small part of my brain

PIECE OF MY BRAIN

PIECE OF MY BRAIN

LETS MAGNIFY IT MORE

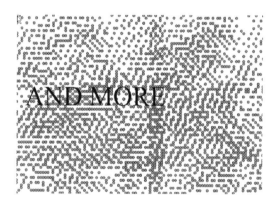

AND MORE

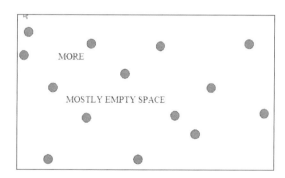

MORE

MOSTLY EMPTY SPACE

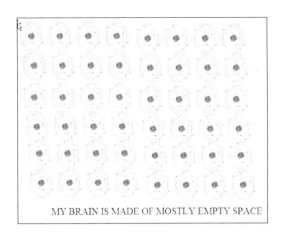

And so is yours!!!!

And your parents….

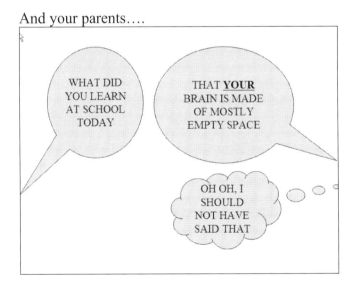

There is a very cool simulation of all of this at
http://micro.magnet.fsu.edu/primer/java/scienceopticsu/powersof10/
It will take you from billions of miles away from earth down to the parts of an atom.
Take a look.

The other thing about atoms (and all molecules) is that they move

Atoms are not living things, they are not alive, but they still move all the time. They make living things, in fact they make everything, but they are not living themselves. They move because they have energy. Take a look at this cool simulation from Phet, where you can change the energy in a group of molecules, see what happens, https://phet.colorado.edu/en/simulation/gas-properties

Atoms (and molecules) move, not necessarily from one place to another but they at least shake and vibrate at *all* times. They are always in motion. This is actually from a type of energy called **thermal energy**. Heat makes them vibrate more; less heat makes them vibrate less. At the temperature of absolute zero (when there is no heat left) they would stop, but this does not happen normally, the temperature of absolute zero is – 459.67°F.

I might mention at this point that there is no such thing as cold in science, only the absence of heat (or more properly thermal energy). So you can never let the cold into your house in the winter, but you can let the heat out!

You are familiar with **solids**. The atoms in solids *move a little bit*, they shake and vibrate, but stay in position. They do not change shape. It is kind of like when you are in a super crowded hallway trying to get to class, or better yet, when you are in the back seat of a car with too many other people. You are moving but not going anywhere.

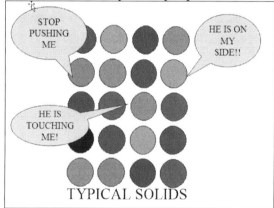

Liquids move even more, they *slide around* in a big group. This is why liquids are always changing shape when you spill them. It is kind of like when the hall is not quite so crowded and everyone is rushing to class and sliding between each other. Or better yet, Wallmart on black Friday when the doors open. It is super crowded but you do move to a new location, but there is a lot of bumping going on.

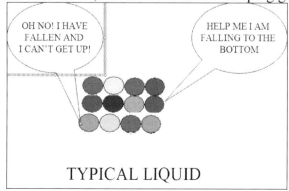

76

Gases move the *fastest*. They can fill the room in seconds. If you go to class after eating a lot of beans, you know this is true. They fly around. It is kind of like when the halls are not so jammed with people and a school lock down is announced. Everyone runs for a classroom.

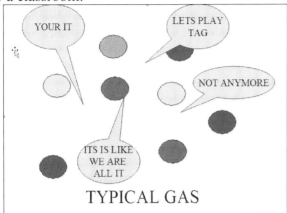

TYPICAL GAS

There is a really cool (and accurate) simulation of how matter changes state at, https://phet.colorado.edu/en/simulation/states-of-matter-basics anther amazing Phet simulation.

Molecules and compounds

At this point I should probably tell you what happens when you put atoms together. Although atoms are the building blocks of matter, very few are found alone. They join with other atoms, to form molecules. When atoms join into molecules they are said to be **chemically combined**. They attach to each other with what are called chemical bonds (a form of energy) that locks them together. They do not let go easily. Molecules that are made of more than one *kind* of atom, like water (H_2O), are called **compounds.**

By the way the chemical bonds that hold atoms together are super strong; you cannot break them with force. No matter how many times you hit a water molecule with an axe, it will not separate the oxygen from the hydrogen, so if you decide to get into your pool and start smashing the water with a baseball bat, you will not break even one chemical bond, you will just get wet.

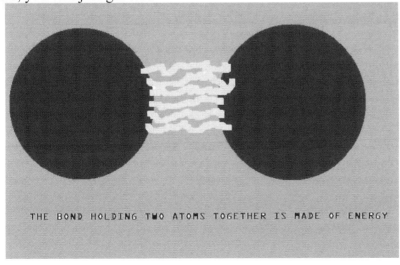

THE BOND HOLDING TWO ATOMS TOGETHER IS MADE OF ENERGY

Yet another fun simulation from Phet is how to make your own molecules, https://phet.colorado.edu/en/simulation/build-a-molecule. Have fun, play, which is actually science.

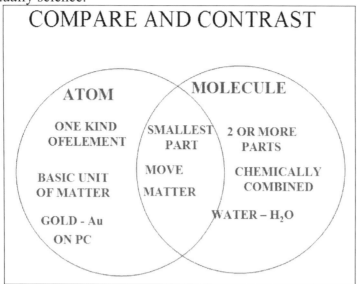

A day in the life of Earwig Hickson III

Today I got my "Young scientists build your own universe kit." It came with 3 boxes. One filled with protons, one with neutrons and one with electrons.

The first thing I did was make all 92 kinds of atoms (elements) and set them aside. I made a factory that squirted out protons, neutrons and electrons.

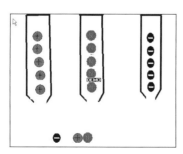 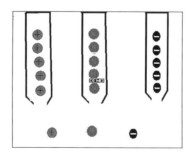

The protons neutrons and electrons made all the kinds of atoms.
https://www.youtube.com/watch?v=EMDrb2LqL7E

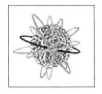 

Then I started putting the atoms together into compounds and molecules

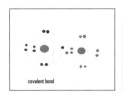 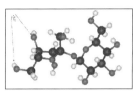

Then I mixed them together to make all kinds of stuff

I made my own universe!

Unfortunately as my universe got bigger it kind of ate the old universe. My new Earth replaced the old one and things went down hill from there. On a brighter note my universe was a much nicer place with no wars and bad stuff, since I got to make the rules! I still got grounded because I used some electrons from my Dad's 1986 BMW R80RT motorcycle when I was a few short. I should have got them from his paddle.

It is hard being a genius, but I am starting to like my corner, the little spider that lives there, Tom, is getting big.

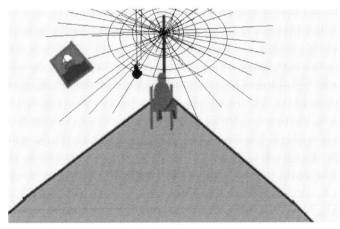

I spent the time in my corner figuring out how I could tear down my Dad's other motorcycle into protons, neutrons, and electrons. I would use them to make a new …….

Review of terms – Quizlet:
https://Quizlet.com/124255014/chapter-7-volume-one-atoms-flash-cards/?new

Fun things to Google

J. J. Thomas
Ernest Rutherford
James Chadwick
Build atoms yourself

Links

Atoms by Lincoln - https://www.youtube.com/watch?v=HDgQ8FgDJVk

Just how small is an atom TEDed – an excellent video that describes just how small an atom is.
http://ed.ted.com/lessons/just-how-small-is-an-atom

From far away to very close – An extremely cool animation starting from outside our galaxy and ending up in an atom
http://micro.magnet.fsu.edu/primer/java/scienceopticsu/powersof10/

How do we know that atoms are real? A short film by Ted-ed.
http://ed.ted.com/lessons/the-2-400-year-search-for-the-atom-theresa-doud

TED-ed Periodic table of the Elements – This shows very cool movies about each Element. Some good explosions too.
http://ed.ted.com/periodic-videos

How do we know what the universe is made of? A film from Ted-ed.
http://ed.ted.com/lessons/what-light-can-teach-us-about-the-universe-pete-edwards

Large Hadron Collider, what does it do? How does it work? It is very advanced science. A film by Ted-ed. http://ed.ted.com/lessons/brian-cox-on-cern-s-supercollider

Phet – make your own atoms – fun and educational
https://phet.colorado.edu/en/simulation/build-an-atom

Atomic Structure: https://www.youtube.com/watch?v=sRPejoNktKE

Chapter 8
Combining matter into new stuff

Movie of the power point for this chapter:
https://www.youtube.com/watch?v=0qhOSn3aMWg

The wonders of science

It was in 1939 when Wallace Caruthers while working for Dupont developed a man made fiber that would change the world, nylon. Originally a substitute for silk it led to the development of plastics. Plastics could be made into any shape imaginable, suddenly clothing became inexpensive, food containers became sterile and toys became safer. Nylon ushered in an age of disposable inexpensive products, the age of polymers, the age of synthetics, the age of substitutions.

How nylon changed the world, http://www.smithsonianmag.com/smithsonian-institution/how-75-years-ago-nylon-stockings-changed-world-180955219/?no-ist

What you know

So far you know that the building blocks of all atoms are *protons, neutrons and electrons*. These come together to build the 92 or so basic units of matter called atoms, such as gold (Au) or oxygen (O). Each kind of atom is also a different element, the names on the *periodic table of the elements*, and these make everything you see. Some of these things are molecules, which is when the atoms are attached by a chemical bond, and will not break apart very easily, such as oxygen gas (O_2). Some of these are compounds, which are simply, molecules made of *two or more kinds of atoms* such as water (H_2O). The last group are mixtures, which just as it sounds are at least two different things mixed up. Like chocolate and milk. I think it is best to just go through them all, even though you are supposed to know some of them, but if your brain is like mine you forget sometimes.

Elements

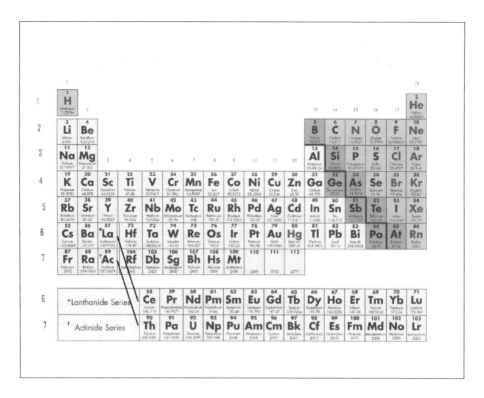

Remember elements are found on the periodic table and the *smallest part is an atom*. This means that all elements are considered *pure substances*, in other words not mixed with anything. A chunk of gold is a pure gold because there are no other *kinds* of atoms attached to the gold atoms.

Most elements can chemically combine with other elements to make molecules. If the element carbon (C) combines only with other carbon atoms, it is still pure Carbon; it is just a big molecule of carbon. It is still a pure substance. Diamonds and the graphite in your pencil are pure carbon, just in different molecular forms.

These are some of the common molecules made of just one kind of atom (or element). They are still pure elements.

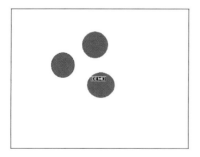

Helium atoms (He) in a group are pure helium. Helium is one of the few elements that will not bond with any other atom. They act a little snobby and are never found as a molecule, only alone by themselves.

Gold ring – all gold (Au)
Diamond – all carbon (C)

82

O_2 – oxygen gas is all oxygen (O)
H_2 – hydrogen gas is all hydrogen (H)

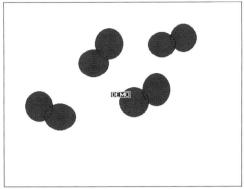

Oxygen molecules (O_2) are still pure oxygen. Oxygen loves to form chemical bonds and is never found as a single atom, only in molecules.

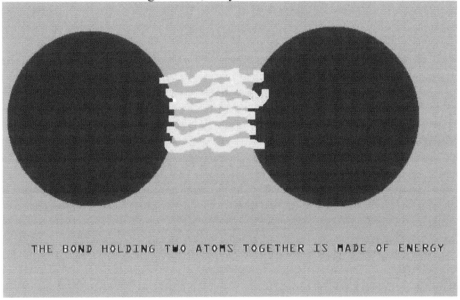

THE BOND HOLDING TWO ATOMS TOGETHER IS MADE OF ENERGY

Compounds

Compounds are molecules but instead of only being made of one kind of atom they have *two or more different kinds of atoms attached with a chemical bond*. There are millions of compounds and they are *never found on the Periodic Table*. Water (H_2O) is a compound since it is made of hydrogen and oxygen bonded together. You will never find water on the Periodic Table, unless you spill some on it. By the way, these are pure substances too, just not pure elements, they are pure compounds. So you can have pure water, if water is all you have.

One thing about a compound is that no matter how you try, you cannot separate the atoms by breaking, filtering, cutting or smashing the compound. These are called *physical methods* of separating materials. You can filter pollution out of water but you can never cut the hydrogen off the oxygen. Not with physical methods. You can by using energy but this is a chemical method, which does not count (yet). You can smash a pool of water with a baseball bat all you want, or you can chop it with an axe all you want but you will still have pure H_2O and you will just get wet. Compounds are held together very strongly by those chemical bonds.

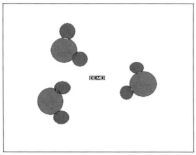

Water molecules (H2O) are made of 2 atoms of hydrogen and 1 atom of oxygen chemically combined (attached).

Some common compounds you have heard of and assorted models to show how they are bonded together.

Water (H_2O)

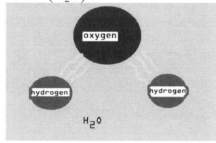

Table salt (NaCl) – one atom of sodium combined with one atom of chlorine

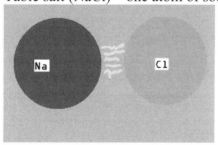

Ammonia (NH_3) – one atom of nitrogen combined with 3 atoms of hydrogen

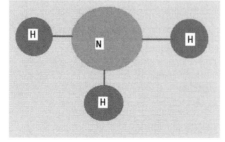

Carbon dioxide (CO_2) – 1 atom of carbon combined with 2 atoms of oxygen

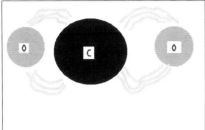

Sugar- glucose ($C_6H_{12}O_6$) – 6 atoms of carbon combined with 12 atoms of hydrogen and 6 atoms of oxygen

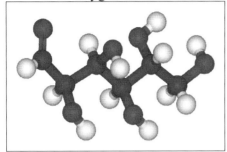

These are compounds because they are made of more than one kind of atom

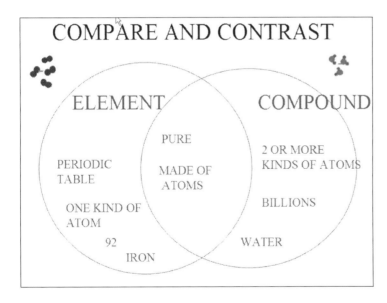

COMPARE AND CONTRAST

ELEMENT

COMPOUND

PURE

MADE OF ATOMS

PERIODIC TABLE

2 OR MORE KINDS OF ATOMS

ONE KIND OF ATOM

BILLIONS

92

WATER

IRON

Mixtures

This brings us to the biggest group of all. Nearly everything you see is some kind of mixture. A mixture is *not a pure substance*, that's what mixture means, more than one substance mixed together. The secret to understanding a mixture is to realize that it is not *chemically combined* like a compound is. There is no attachment between the parts. The parts may be next to each other or even touching but are not attached in any way, they can separate easily. It is kind of like a salad; the tomatoes are not attached to the lettuce and can easily be removed if you do not like them. There is no compound called tomato-lettuce.

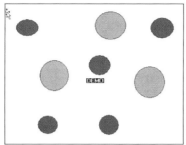

A mixture of 2 kinds of atoms

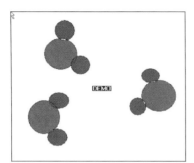

A compound of oxygen and hydrogen (water) notice the difference?

The thing about mixtures is that since they are not bonded together they can easily be separated by what we call **physical methods**. Physical methods are things like filters, magnets or even just letting them settle out on their own by gravity. Mixtures can be separated without using energy or any chemical methods.

We separate mixtures in our society all the time. The water you drink has been filtered to get all the yucky stuff out. A fish aquarium or swimming pool has a filter to remove the fish or people dirt. Rocks settle out at the bottom of a river. I always pick out the marshmallow charms in my Lucky Charms and eat them last. Does anyone else remove the green beans from beef stew? Why do people even put them in there? We wash our clothes to remove the dirt from the cloth. When you clean your locker or worse yet your room, you are separating a messy mixture.

Physical methods of separating a mixture

Using magnets to separate iron from trash at a garbage processing plant is one way. Since Iron is magnetic and old spaghetti is not, the mixture separates and the iron can be sold and recycled. If you want to try something really disturbing do this; after you pour the milk on your iron-fortified cereal, mix it up good then put a magnet at the bottom. You will find small bits of metal (iron) stuck to the magnet. Surprise, there is iron in your cereal. By the way this is not a bad thing, they put that iron in for our health, and it is a vitamin.

Filters are able to remove most things from water by passing the dirty water through tiny holes. The small water molecules go through but the chunky stuff gets trapped. This is how your fish or pool filters works. Sand, charcoal and different materials are used to do this. https://www.youtube.com/watch?v=YaOOmaGzx0g. By the way this will not remove other tiny molecules like salt (NaCl); it will not work if you want to drink

seawater. For that you need *distillation*.
https://www.youtube.com/watch?v=00kKPOs_FA4

The density of materials can be used to separate them too. Less dense things float, more dense things sink. If you put a mixture of sand and saw dust in water, the sand will sink and the wood will float. This is the idea behind panning for gold. You scoop up some mud that you think might have gold in it with a pan. You swirl water in the mud and all the less dense stuff spills out leaving only the denser material. Gold is rather dense so it should stay.

Chromatography is when you let a mixture separate by capillary motion. It is fun to write a circle around the center of a paper towel with a dark *water based* marker, then put the unmarked end of the towel in a cup of water. The water will soak up the towel and when it hits the marker, all kinds of colors will separate out. It is cool.

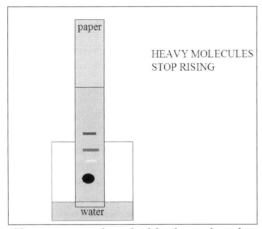

Chromatography of a black marker dot

Distillation is when you let water evaporate or boil out of a dirty mixture. If you boil sea water into water vapor, the salt is left behind, if you catch the water vapor and condense it back into water, the water is pure. If you let sugar water evaporate away, only the water leaves. What is left behind is rock candy!

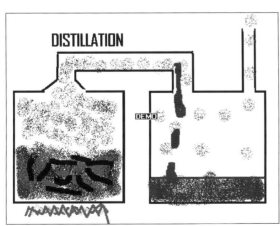

Distillation of polluted water into pure water.

Some things you can just leave alone and it will settle out all by itself (with some help from gravity). Some salad dressings (Italian) that are made of oil and vinegar, settle out quickly In fact you have to shake them before you use them, or you will get all vinegar on your salad. If you do not drink your chocolate milk for a while, most of the chocolate will sink to the bottom.

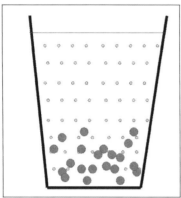

Chocolate milk I forgot to drink – yuck

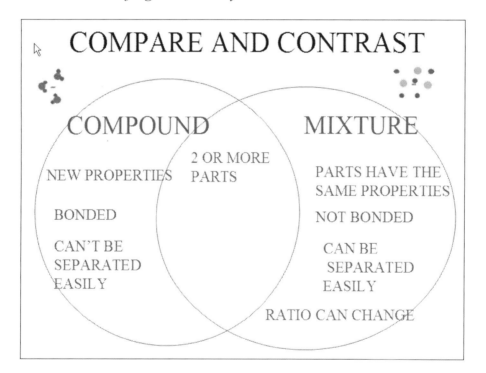

Two groups of mixtures

There are two main groups of mixtures, the ones that stay mixed and the kind that does not stay mixed. *Homogeneous* mixtures stay mixed and you do not have to shake them when you use them. *Heterogeneous* mixtures settle out and have to be remixed all the time. This is why chocolate milk needs to be shaken before drinking but soda does not.

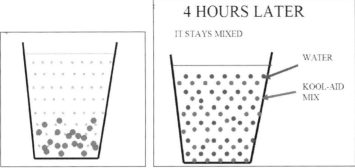

A heterogeneous mixture of chocolate milk left sitting for 4 hours compared to a homogeneous mixture of kool-aid

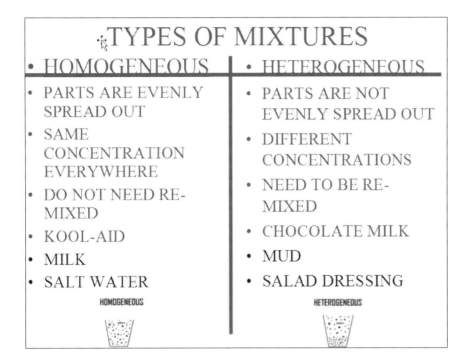

Review

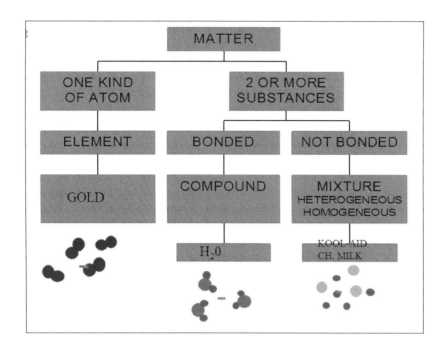

A day in the life of Earwig Hickson III

Today I decided I wanted to become a great chef. I knew just what to do, find all my favorite foods and mix them together! All good things mixed together would no doubt be great. Why is it that no one had thought of this before?

I started with my favorite food, BACON! I put it in the blender and ground it up. My second favorite food is chocolate so in it went, the mixture that resulted did not look that good but taste would be everything. I like jelly so that went in too, and I did not forget jelly beans either, mostly the green ones. I was getting excited as I added a left over pork chop and some shrimp. Candy is always good so I added some taffy and some more chocolate. Peanut butter goes well with chocolate so I added that. I almost forgot orange juice, but didn't, and milk. The mixture looked very uniform so I knew it was homogeneous, I could not see through it so it was not a solution. When I added a pancake it only got better, especially after the syrup. I knew it was not finished so I found some marshmallows because they go with chocolate and some ketchup because it goes with everything. I found some cheese in the back of the fridge and blended it in. Crackers go with cheese, pizza has cheese and so do tacos, and I added them. The blender was starting to have trouble keeping up, but improved after I added some honey and soda. Corn! I had to add that along with some spaghetti and a hot dog. My masterpiece was nearly finished.

I looked at my mixture. All my favorite flavors combined. Since it was a mixture all the original ingredients were still there. The hot-dog, the corn, the cheese had all donated their flavors to my concoction. But when I tasted it, it was not quite right. I could still taste every ingredient since a mixture still has all the original properties. I put some red food color in to improve the looks, and some vanilla to improve the smell. As hard as it is to believe, the taste was just not right. What was I missing?

I realized that I had not done anything special by just mixing stuff together. It needed some energy, because energy causes molecules to bond together into new compounds. That was what I was after, so I put it into the oven to cook. When the heat was absorbed the molecules got more energy, chemical bonds were broken and reattached into new compounds that may have never been seen on earth before! These compounds with their new properties were nothing like the original ingredients. The smell was new, the color had changed to a new dark tan, and the consistency was much thicker. I had invented the new compound: *baconchocolatejellybeanporkchoptaffy-peanutbutterorangejuicemilkpancakemarshmallowketchup-cheesecrackerspizzatocohoneysodacornspaghettihotdogvinilla*. It tasted like, um, not what I expected, evidently as the bonds broke and reattached something changed. The cheese was not cheese anymore, the bacon was not bacon anymore and the ketchup was just not the same. The new compound had new properties and a new taste.

I realized in hindsight that compounds are totally different from their parts. Chemically combining all these things changed the properties, everything was chemically combined. It had a new color, a new odor, and unfortunately a new taste. I added a whole lot of wintergreen mints to cover the smell and some blue food dye, which just made the *stuff*, as I called it now, a deep brown. At least it smelled better. I put a topping of whipped cream and a maraschino cherry on top. I put it in the fridge.

Now I had planned to try and fix it the next day. I was going to try distillation to separate some of the less desirable parts by boiling and condensing, and maybe deep-frying it. Unfortunately, in the middle of the night my dad got hungry. In his half awake state he thought he saw a nice chocolate mousse sitting on the top shelf of the fridge. He ate the whole thing.

The bathroom was occupied the next morning, with strange sounds coming from inside, and I suspected why. I did not complain, or question, and hurried to the bus stop, before he came out. He missed two days of work and his bowling league. He was grouchy.

Review of terms – Quizlet:
https://Quizlet.com/124255852/chapter-8-volume-one-combining-matter-flash-cards/?new

Fun things to google

Interactive Periodic Table of the elements
How is trash recycled?
Cornstarch and water mixture
How to purify water
Chromatography
How to make rock candy
How to make crystals

Links:

Build molecules Phet – a fun simulation
https://phet.colorado.edu/en/simulation/build-a-molecule

How drinking water is made
https://www.youtube.com/watch?v=KMP9-49I1U4

A wastewater treatment plant simulation
https://www.classzone.com/books/ml_science_share/vis_sim/ewm05_pg52_treatment/ewm05_pg52_treatment.swf

The chemistry of making a cookie. A film from Ted-ed. http://ed.ted.com/lessons/the-chemistry-of-cookies-stephanie-warren

The science of macaroni salad. A film by Ted-ed. http://ed.ted.com/lessons/the-science-of-macaroni-salad-what-s-in-a-mixture-josh-kurz

How to make a wave bottle (density bottle).
https://www.youtube.com/watch?v=_HSsu11-frQ

Chapter 9
The common states of matter

Movie of the power point for this chapter:
https://www.youtube.com/watch?v=7nkK5zatonA

The wonders of science

On October 28, 1959, thousands of people in United States witnessed a frightening display of flashing lights high in space coming toward Earth. Not unlike a science fiction movie about alien invasions, telephone calls were made by worried witnesses to the police. It was not an alien invasion or UFO's of any type; it was Shotput 1, an experiment by NASA, testing the idea of putting a giant (100 feet in diameter) Mylar balloon in space. The idea was to make a satellite that could bounce radio waves (microwaves actually) around the world (like a giant mirror) for communications. The Giant balloon was folded and put on top of a rocket. Inside the balloon, there were solids and liquids, which would turn into gas when the payload was placed in orbit around the earth. Unfortunately there was also a little bit of air. The launch worked perfectly but when the balloon began to inflate the small amount of air expanded so rapidly (since space has no pressure to hold it together) that the whole balloon popped and shredded into thousands of reflective pieces, which fell slowly back to earth reflecting sunlight as they fell. It was quite a light show. NASA learned from that mistake and in 1960 successfully put a balloon satellite (ECHO 1) into space. This ushered in the age of satellites, which we now depend on for instant communications (TV, Phones and so on), weather prediction, climate research and even GPS navigation. A failed experiment like this one is not a failure at all, just a step closer to success.

More on the exploding balloon satellite, http://history.nasa.gov/SP-4308/ch6.htm

Project ECHO, https://en.wikipedia.org/wiki/Project_Echo

The common states of matter

The 3 common states (or phases) of matter are *solid, liquid and gas*. There are others but for our purposes these are good enough. You should already know that solids move slowly; they just kind of shake and vibrate. Liquids move a bit more by sliding among themselves. Gases really fly. But there is more.

Solids

By definition *solids have a definite shape and volume.* They do not change much. Oh, they can shrink a bit when it gets cold or expand a bit when they get hot, but for the most part they do not change their volume. A brick you saw on Thursday looks the same on Saturday. You also cannot compress a solid, because the molecules are already so close together you can't crush them any closer. Try squeezing a brick smaller, it just does not work well, there is not much empty space to squeeze out. Its shape does not change either. If you spill a block of lead you still have a block of lead lying on the floor, and it looks the same. If you drop a glass and it shatters, the pieces may be smaller but you could put it back together and it would be the same shape as before, so solids do not change volume or shape in any way. They also hurt when they hit you. https://www.youtube.com/watch?v=7eXxkGP3fp8

Liquids

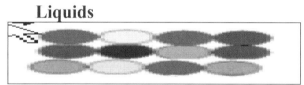

https://www.youtube.com/watch?v=EbXJVs6MulE

Liquids are different. By definition *they have a definite volume but <u>no</u> definite shape.* They change shape all the time. They take the shape of whatever container they are in. If you spill a liquid it takes the shape of the floor. If you put them in a cup they are shaped like a cup, you can't do that with a solid! Their volume on the other hand does not change, because the molecules are already very close together. If you take a hand full of water and try to squeeze it, water will just squirt out between your fingers. The volume can change a little though, as the temperature changes, just like solids. Cold liquids will get a bit smaller, heat them up and they expand a bit, but not much. So for our purposes liquids do not change volume. They do not hurt much when they hit you, they slosh around you.

The one odd liquid that messes this idea up is water. Water is special, it shrinks when it gets cold but only until just before it freezes, then it goes against all common sense and expands. This is why water pipes burst when they freeze, and why you need to drain your garden hose before winter, but only water acts like this, everything else keeps getting smaller. To prove this to yourself put a completely full container (a cup or something) of water in the freezer, it will be bigger than the container when it freezes, a closed plastic water bottle (completely full) might burst, since water expands when it freezes.

Want to see why pipes freeze in the winter? Try this. https://www.youtube.com/watch?v=HFMJp2xaKCk

Want to see a cool experiment you can do at home? How about an exploding soda bottle, resulting in soda-snow? https://www.youtube.com/watch?v=WFyaL6iozKY

Gases

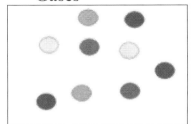

Gases are the freest of them all; they pretty much do what ever they want. By definition *they do not have a definite volume or shape*. They have a tendency to expand and compress very easily to fill whatever container they are in. The molecules are already pretty far apart so they can get closer, or farther if they feel like it. With a bike tire pump you can take a whole room full of air and squish it into a bike tire. If the temperature increases they get bigger, a lot bigger. They always take the shape of the container you put them in. In your classroom the gas (air) is in the shape of the classroom, there is no corner that has no air in it, it is everywhere. When you breathe, the air takes the shape of your lungs. Gases fill every nook and cranny. Sometimes this is good, sometimes this is bad.

https://www.youtube.com/watch?v=csm6UcpmfBc

If you blow up a party balloon in your house and put it outside in the cold it will shrink, the molecules are getting closer together, bring it back in the warm house and it will get big again as the molecules spread out. Put it in a hot place, like over a heater, it will get bigger yet. This is not because there is more air in it, it is because the air inside is spreading out, and it has expanded. https://www.youtube.com/watch?v=iSxBukT2hok. https://www.youtube.com/watch?v=RGTMIcAh4KM

When a skunk is angry and sprays its gas, it quickly fills the container it is in. It is best not to let an angry skunk into your house. You will regret it. When the girl across the room thinks it is necessary to squirt her nasty perfume, the whole room is filled, same with the kid that had beans for lunch or the kid who never takes a shower. Gases expand to fill the container they are in, and they fill it completely. https://www.youtube.com/watch?v=VZwlKANA43w

Remember in the first JAWS movie when the hero shot the scuba tank and all the air expanded the shark until it popped? Although not very realistic, the idea is correct. He should have made him swallow the tank first though.
https://www.youtube.com/watch?v=FpxOLhuNXfM

By the way the reason you "think" you can smell a solid or liquid is because of gases it is producing. These gases you can smell. So when your little sister has a dirty diaper, it is not the solid you smell but the liquids evaporating from the mess into gases. Gases get into your nose. . https://www.youtube.com/watch?v=O414BHvy4n4

Review of the 3 common states of matter
COMPARE AND CONTRAST

	SOLID	LIQUID	GAS
volume	FIXED	FIXED	CHANGING
shape	CONSTANT	CHANGING	CHANGING
Particles distance	CLOSE TOGETHER	CLOSE TOGETHER	FAR APART
Particles speed	VIBRATE	SLIDE AROUND	FLY AROUND
compress	NO	NO	YES

A day in the life of Earwig Hickson III

I finally caught that cute black and white, smelly creature that gave me such a bad time earlier. It turned out to be a skunk, a type of weasel. I looked it up on the Internet and it turns out they do make good pets. I snuck him into my bedroom and named him Kitty. I decided to observe before doing any experiments this time.

After a while I noticed he produced a solid from his backside, it did not smell a bit. My research told me that you cannot smell a solid because none of the molecules actually get to your nose where you would detect them, it was also very cold that day, which may have helped, and no gases could evaporate. Imagine that, you cannot smell a solid, only the gases coming from it. That explains why my socks smell worse on a hot day than a cold day, quicker evaporation. It also explains why I cannot smell a brick.

Later my skunk friend made a liquid. The same results occurred; I could not smell it, because all the molecules were still in the liquid, and not in my nose. A hot day may have been different and allowed the liquid to evaporate into a gas, but it was far too cold for that. That must be why water has no odor.

I figured it was safe to do some experiments so I snuck up on my skunk in my scariest Halloween costume (a mask of the periodic Table of the elements – it scares middle school students to death) and screamed TEST. Well this time he produced a gas! And I could smell that, it filled every part of my lungs and nose! It filled every part of the room! It filled every part of the house. The molecules found their way into the noses of my Dad, Mom, and sister. Our eyes teared up. Everyone gagged. My Dad jumped right out the window. My Mom said, "Oh dear," almost a swear word for her. Gases it turns out, you can smell, they go right into your nose, and fill ever nook and cranny.

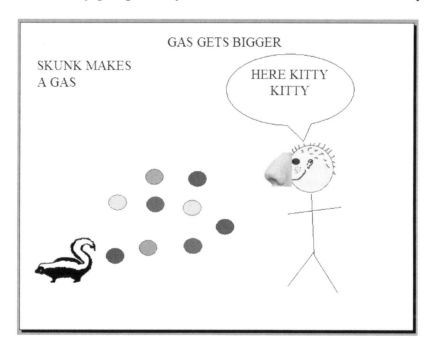

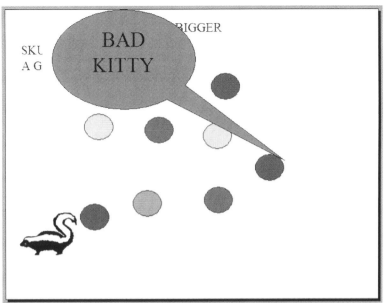

My next few nights in the tent in the back yard were not at all unpleasant. It was also quite easy to delete my Dad's Craigslist ad for a small boy, free to *any* home. It was cold in the tent and the gas did not move very fast. I got used to it and it turns out a skunk makes a very cuddly pet, and no one; I mean no one bothers me when I walk him. He is better protection than a big dog. I tell everyone he is my attack skunk, they take no chances. I love Kitty my skunk. Bullies do not pick on me!

Review of terms – Quizlet:
https://Quizlet.com/124257577/chapter-9-volume-one-common-states-of-matter-flash-cards/?new

Fun things to google

Pet skunks
How do you smell stuff?
Why does poop smell?

Links

Why we can only smell gases. https://www.youtube.com/watch?v=O414BHvy4n4

How do we smell? A Ted-ed Film. http://ed.ted.com/lessons/how-do-we-smell-rose-eveleth

Phet, the states of matter. A fun simulation. https://phet.colorado.edu/en/simulation/states-of-matter

What is plasma? A short film by Ted-ed about the 3 common states of matter and one more, Plasma. http://ed.ted.com/lessons/solid-liquid-gas-and-plasma-michael-murillo

What is a solid? https://www.youtube.com/watch?v=7eXxkGP3fp8

What is a liquid? https://www.youtube.com/watch?v=EbXJVs6MulE

Watch a steel pipe explode when frozen. https://www.youtube.com/watch?v=HFMJp2xaKCk

How to make a frozen soda bottle explode! https://www.youtube.com/watch?v=WFyaL6iozKY

What is a gas? https://www.youtube.com/watch?v=HFMJp2xaKCk

Unit 3
The Properties of Matter

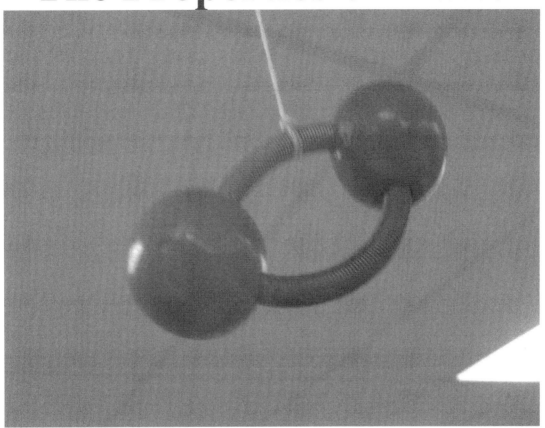

Chapter 10
Properties of matter

Movie of the power point for this chapter:
https://www.youtube.com/watch?v=CxnF9VUA5zw

The wonders of science
In early 1848 gold was discovered in California. By 1849 thousands of people flocked to the area to make their fortune. The California gold rush was on. Finding gold is not easy and when it is found not everyone acts rationally, gold fever is when you get so excited that all you want to do is find more. But how do you know it is gold and not some other mineral that just looks like gold? Many people were fooled by iron pyrite (fools gold) or even purposely faked gold nuggets. It was a wild time.

So how can you tell gold from non-gold? You test it. The first thing you might want to try is hitting with a hammer, real gold will smoosh (it is soft), fools gold or mica will shatter. You might notice that it is very heavy for its size (real gold is very dense). You might scrape it on a ceramic tile (called a streak test), gold will make a bright yellow line, and fools gold will leave a greenish-black line. Gold is not attracted to a magnet. A chemical test often used is to put a drop of Nitric acid on it, nothing happens with gold, but that is not true for the fake stuff. So when it came to identifying gold, science saved the day, again.

More about the California gold rush,
https://en.wikipedia.org/wiki/California_Gold_Rush

Properties of matter
Properties are how you describe or recognize something. How do you recognize your friends? By what they look like, their height, their shape, their hair color, voice, facial features, ugliness, and many other factors. You could actually describe someone in enough detail so someone else could recognize him or her. These are *physical properties*, what things look like, smell like or sound like. You could also describe them by their behavior or how they act; these would be *chemical properties*, what things do.

Physical properties
By definition a *physical property* is a characteristic of a substance that can be observed *without changing* the identity of the substance. It is how you recognize things.

Examples of *physical properties* are how things look. Color is a physical property, odor is a physical property, and so are shape, density, and mass. Other physical properties include things like, is it magnetic, does it dissolve in water, what is its melting point, or its boiling point. A physical property is anything that *describes* a substance to help you recognize it later. Physical properties are always present, it is how you describe and recognize things.

The physical properties of my pet skunk, Kitty, are that he is black and white, furry, smells bad and hisses a lot. Now you know how to identify Kitty, so watch out.

Density

This is a good place to really explain what density is. Remember density is how much mass per unit volume. It was used by Archimedes to identify if a crown was made of pure gold or a mixture of gold and silver. Archimedes used a physical property to identify the material to the detriment of the guy who made the crown. The property of density is how fluffy or compact something is. Which weighs more, a Kilogram of feathers or a kilogram of lead? The obvious answer is they are *the same*. But a better question is which is bigger? Well this is a density question. If they have the same mass, but different volumes the bigger one has a lower density. The feathers are bigger because they are fluffier; it takes a lot of feathers to make a pound.

The crown of Syracuse https://www.youtube.com/watch?v=wEvtahSn_ms

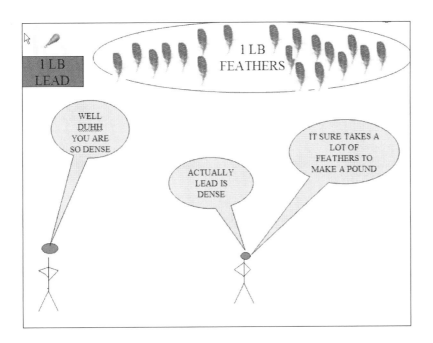

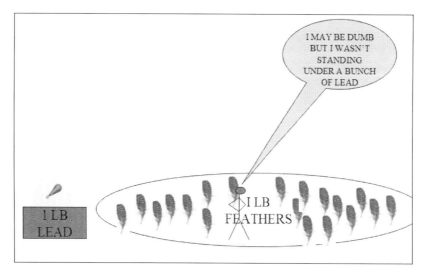

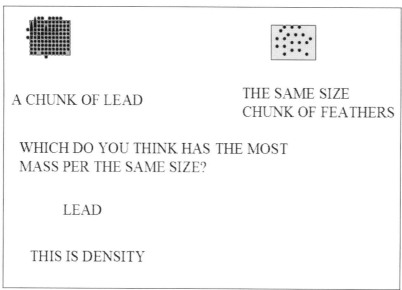

A CHUNK OF LEAD

THE SAME SIZE
CHUNK OF FEATHERS

WHICH DO YOU THINK HAS THE MOST
MASS PER THE SAME SIZE?

LEAD

THIS IS DENSITY

Every now and then there is an oil spill in the ocean. One of the biggest problems for wild life is that oil is less dense than water so it floats on top of water. This is a problem for birds, who think the ocean in made only of water. Birds are denser than oil so they sink in it, but still float on the water, bad news for the birds, who are now covered with oil and can't escape. Nice people often try and save the birds by washing them in detergent, to remove the oil, but this is a big job and they can only save a few. Fish do no better, a large fish chasing smaller fish to the surface get stuck in the oil too. Oil is poisonous to them. Oil spills are a very bad thing for the environment, and we are part of the environment.

The oil spill in the Gulf of Mexico,
https://www.youtube.com/watch?v=8Uax5FRWnvs

You can do a simple density experiment by filling a water bottle half full with cooking oil and half with water, put different objects in and play with it (don't forget the lid).
https://www.youtube.com/watch?v=usKgkrlP0Kw

Physical Change

Remember a physical property is what something *looks like*, a physical description. A physical change is when you do something to the material but do *not* actually change it.

For example you could take a tree, which is made of wood, and chop it into wood chips or saw dust, it may not *look* the same but it is, it is still wood. In addition you could theoretically put the tree back together into the original tree shape, it would be a long useless job, but possible nonetheless. See, you changed the shape but not the actual substance, it is still wood. You started with wood and ended up with wood. In a physical change you always end up with the same *molecules* you started with.

Wood in the shape of a tree and wood in the shape of wood chips (still wood)

You can hit water as much as you want with a bat, you can freeze it, you can boil it, you can drop it off a cliff, but in the end it is still water. You can change the water's shape or size but what you start with is what you end with, at least as far as the molecules are concerned, you still have water. You have the *exact* same material.

By definition a *physical change* is any change to a substance that does *not* actually change the substance, at least not its molecules. Examples include chopping wood (you still have wood), boiling water into steam (you still have water), freezing water into ice (it is still water), and grinding a big piece of iron into iron filings (you still have iron with the same physical properties). Physical changes always result in the *same molecule* with the *same properties*.

When the Hulk in the "Avengers" movie smashed stuff, that was a physical change, all the Hulk does is SMASH.

Chemical property

A chemical property is a *behavior*, how something acts. In the world of chemicals it is how it reacts with other chemicals. In your world a chemical reaction is how you behave with others, do you fight with them, or are you nice, in other words a behavior.

A chemical property describes how a substance can form new substances, it is what things do, how a substance behaves. For example some things rust easily (a slow reaction with oxygen), some burn easily (a fast reaction with oxygen). It is all about how a substance reacts *chemically* with other substances.

Chemical properties are also used to describe a substance. It is useful to know if something is flammable, does it react with air, or other materials. Often in the directions for using household cleaners it says not to mix with specific chemicals. Toilet cleaners should not be mixed with Clorox, oven cleaners should not be placed near aluminum foil (in the trash), these are for safety reasons, and you do not want to burn down your house or poison your cat. Chemical properties are used to describe a substance just like physical properties. By the way some of the most dangerous chemicals you will ever come in contact with are household cleaners, so be careful where you store them. Under the sink is the absolute worst place!

Chemical change

A chemical change is like magic. Something unexpected happens, something that cannot easily be explained, by common sense, an unexpected color, or odor, or heat, or cold, or even a liquid turning into a solid. It is just like magic, but there is no such thing as magic, only science. You can imagine mixing red and yellow and making orange, but what if it made purple? This would be a surprise, and a sign of a chemical change, since it was unexpected.

A chemical change is when you end up with a *new* molecule with *new* properties. The hard part is how do you recognize a chemical change? It is actually rather easy, you end up with something *new*, like a *new* color that you cannot explain, a *new* odor that was not there before. Is there *new* heat that you cannot explain, or *new* "cold" that has no explanation? Did bubbles form that have no reason to form? In a chemical change a new molecule is *always* formed and *new molecules have new properties*. So how do you tell the difference between a chemical and physical change? Try this. https://www.youtube.com/watch?v=qqqmFFCwd7k

When pH test strips are used to measure the pH of a fish aquarium or swimming pool, a change in color represents a *chemical change*. When you put Alka-Seltzer in water and it bubbles, this represents a chemical change because you cannot explain where the bubbles came from (Alka-seltzer is not made of gas). Some reactions get hot, like a campfire, this is a sign of a chemical change. Some reactions get cold for no apparent reason, like a cold pack in a medical kit for a sports team, another sign of a chemical change. Sometimes a new odor is formed, a dead fish smells OK when it is fresh, but after a few days it is not the same, a *new* odor is formed, a chemical change has occurred. In a chemical change a new molecule is always formed. Let's face it; fresh fish is not the same as old rotten fish.

After a chemical change you have a new molecule, with *new properties*. When you burn wood in a campfire you end up with smoke and ashes, these are different from the original wood. You cannot put the ashes back together into a log. You cannot build a house out of ashes; the physical properties are not the same. When iron rusts into iron oxide (rust), it is not the same. A bridge made of iron is much better than one made of rust. These are all examples of a chemical change; the final stuff is just not the same as it was.

Some common ways chemical changes occur are burning (breaking and oxidizing the chemical bonds in wood), electrolysis (breaking the chemical bonds that hold water together), chemical reactions (baking soda and vinegar make carbon dioxide gas), and cooking food (raw eggs and cooked eggs are not the same).

Review

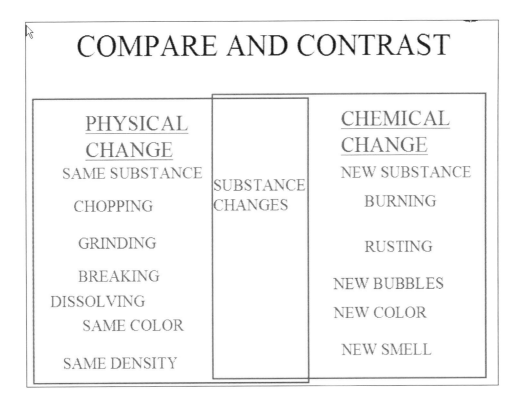

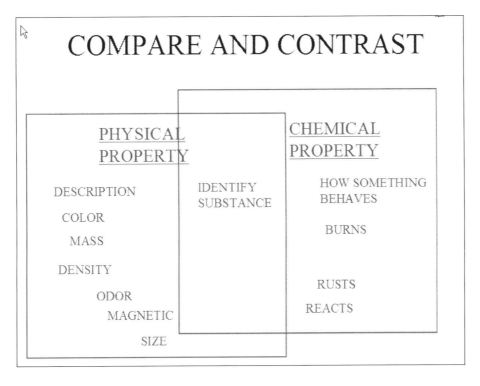

A day in the life of Earwig Hickson III

It was time for bed, but I was thirsty. I had just learned in science about density and I wanted a density (or stacked) drink. I began with three different colors of Kool-aid and added different amounts of sugar to each flavor. The yellow color got no sugar, the blue got the normal amount and the red got twice as much as normal. I also put a nice layer of chocolate syrup in the bottom. I carefully poured the red, then the blue, and finally the yellow, it looked and tasted great!

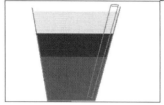

I had a dream. It was a wonderful dream when it started. I was a bird flying free in the sky. I could see everything from up there. It was so peaceful. I was a Duck. My name was Arnold. I was Arnold the duck.

In my dream I decided to fly south to the Gulf of Mexico for the winter, being a bird I could take the ultimate vacation to a tropical paradise. Off I flew. I could not wait to swim in the Gulf of Mexico because I knew that since Ducks are less dense than water, I could easily float, and swimming would be easy. I could even take a nap on the water; I had never done that before.

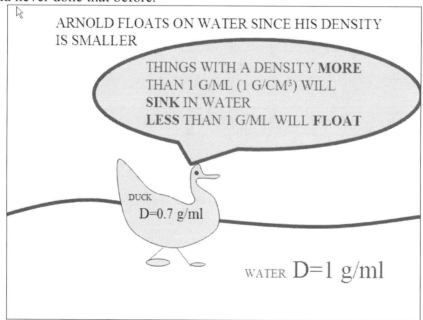

A sad thing happened just before I arrived at my vacation paradise. There was a large oil spill right where I was going to land. As it turns out oil is less dense than water so it floats on the top of water. And ducks? What is the density of a duck? Being a duck in my dream, I did not even know about oil spills, I thought the ocean was made only of water.

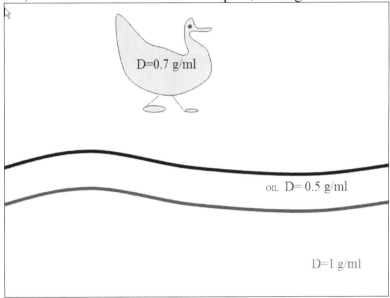

Sadly it turns out that ducks are denser than oil and sink. They still float on water, but that isn't much help if the oil is very deep.

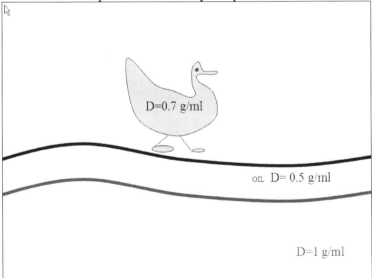

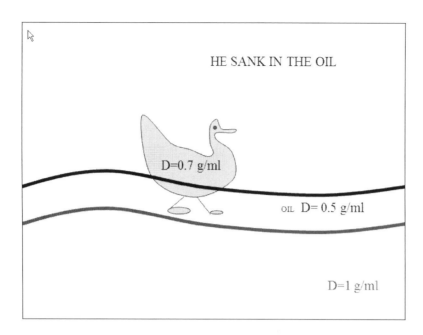

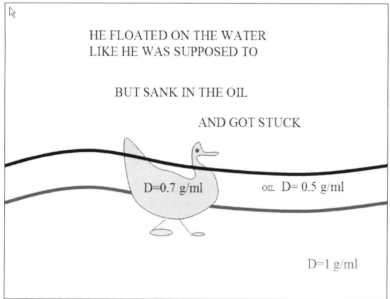

I ended up floating on the more dense water, like ducks are supposed to, but sank in the less dense oil. I was trapped. My vacation was ruined.

I should have woke up at that point but sadly I did not, the nightmare continued. An alien flying saucer came down and scooped me up, water oil and duck with one big spoon. I thought they were going to save me but noooooo; they put me in a big blender and mixed every thing up. The result was a density column of oil, water and duck parts. They were studying density, just my luck.

That is when I woke up. No more special density drinks for me before bedtime.

Review of terms – Quizlet:
https://Quizlet.com/124258195/chapter-10-volume-one-properties-of-matter-flash-cards/

Fun things to Google
Food chemistry
The crown of Syracuse
Deepwater Horizon Oil spill
Mentos experiment
How to make a density column
How to make a wave bottle
Density experiment you can drink

Links
Egg suspended in half water and half salt water
https://www.youtube.com/watch?v=ZgdqJwlG7lQ

Burning metal salts – notice the cool colors:
https://www.youtube.com/watch?v=36ahEfYvA-A

An unexpected chemical change – this is called a clock reaction:
https://www.youtube.com/watch?v=p-vEti8zEnM

Alka-seltzer in a wave bottle
https://www.youtube.com/watch?v=usKgkrlP0Kw

How to separate beads using density:
https://www.youtube.com/watch?v=AS5xUhNkZhI

How to make a density column. https://www.youtube.com/watch?v=-CDkJuo_LYs

Chapter 11
Changing phases (states) of matter

Movie of the power point for this chapter:
https://www.youtube.com/watch?v=NPQuLnqRxnQ

The wonders of science

In Death Valley rocks seem to move by themselves, leaving sliding marks in the dirt as if someone or something pushed them. The trails can be 70 feet long and the rocks can weigh 600 pounds. What kind of magic is this? Not magic, only science. For years it was a mystery how heavy rocks could "walk" or "slide" by themselves in the middle of a desert. Is this one of those mysteries that is beyond human understanding? Is it beyond science? Not so fast, scientists may not know how something works but that does not mean we never will. We just don't know, YET. It turns out some guys did figure it out with a long boring experiment. They actually watched for two years until the rocks moved, and when they did, the mystery was solved. It was ice.

Walking rocks, http://voices.nationalgeographic.com/2014/08/27/watch-death-valleys-rocks-walk-before-your-eyes/

Things to remember

Remember the **common states (or phases) of matter are solids, liquids, and gases**. The only real difference between them is the **amount of thermal (heat) energy they have**. Thermal energy makes molecules move faster, and this can cause the matter to change form (or phase). There is a very simple simulation for this at http://www.bgfl.org/bgfl/custom/resources_ftp/client_ftp/ks3/science/changing_matter/index.htm. Notice it is all about *heat* being added or taken away.

Walking on water

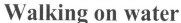

The first time I walked on water I had no idea it was special. I just did it. I did not do anything special to prepare for it. I did not tell anyone I was going to do it. I took no safety precautions. I thought anyone could do it. Later I heard that walking on water was not something normal people were supposed to be able to do. I began to think I was special, maybe an alien from another planet or a wizard. For years I kept my power a secret. I lived a lie.

The first time I walked on water, it was on a very deep lake called *Lake Wet*. The hardest part was walking through the snowdrifts to get to the lake and keeping warm in the super cold wind. Later I realized that most people do not think this counts. They called what I walked on "ice", no big deal. But why shouldn't it count? Is ice not made of water? True I could not do it in the summer, but ice is still water.

Matter can change from one state to another

OK, so I have no magical powers, anyone can walk on ice. The problem is how do you define water? Is it a liquid? Sometimes it is. Is it a solid? It is that sometimes also. It can also be a gas. These are called the *phases or states* of water, but they are all water or H_2O. In fact nearly all matter can be in *any phase* depending on the conditions (temperature and pressure). Water is the easiest to understand because we see all its phases on earth. That is because the temperatures on our planet vary from darn cold (or more accurately a *lack of heat*) to horribly hot.

Do you know the recipe for making ice? You put it in the freezer and the next thing you know is you have ice. But why? What is the difference between ice and water? The reality is: not much. Granted throwing a gallon of water on someone is not the same as throwing a block of ice on him or her, but that is about it. The real difference is the heat energy (actually thermal energy – more on that later) inside. Liquid water has *more energy* than solid water. That is correct; I said "solid" water that is what ice really is. It is the water you chew.

When you put the *liquid water* in the freezer, it had a lot of energy in it. The molecules were moving so fast they slid around each other. They were always moving. The freezer pulled that energy out of the water (and out the back of the fridge). This caused the molecules to slow down until they got stuck in place. The liquid turned to solid. If you want to really get fancy, put the ice cube (*solid water*) into a hot oven and it will turn back into *liquid water*! Heck if you keep heating the water it will evaporate into *gas water*! This is because as you add energy the molecules go faster until they fly away into water vapor.

You can play around with this idea at Phet – changing states of matter - https://phet.colorado.edu/en/simulation/states-of-matter

https://phet.colorado.edu/en/simulation/states-of-matter-basics

So now we see that ice water and steam are all made of the same molecule, H_2O. The only difference is how much energy the molecules have, thus how fast they are moving. You can easily change the phase of matter by adding heat or taking heat away (by the way there is no such thing as cold so you cannot add cold!) Now for the bad news, each of the changes of phase has a name, time for some vocabulary!

MELTING	SOLID TO LIQUID
FREEZING	LIQUID TO SOLID
EVAPORATION	LIQUID TO GAS (SLOW)
SUBLIMATION	SOLID TO GAS
CONDENSATION	GAS TO LIQUID
BOILING	LIQUID TO GAS (WITH BUBBLES)
BOILING POINT	TEMP WHEN A LIQUID BOILS
FREEZING POINT	TEMP WHEN A LIQUID FREEZES

Most of these words are familiar; the new one is **sublimation**, where a solid turns directly to a gas. You have seen this but did not know it. It is kind of like evaporation except one molecule of the solid flies off into the air at a time. Did you ever notice that snow in your front yard disappears even when the temperature is below freezing? Dry ice is *solid carbon dioxide* (CO_2) but it will not melt into liquid, it sublimates and goes off into the air. By the way you might see white fog rising from the dry ice but it is actually freezing water vapor. Did you ever notice that the size of an ice cube in a frost-free freezer shrinks over time? Did you ever put snowballs in the freezer during wintertime only to go get them in the summer and they are very small? This was sublimation, the ice turned into gas. Frost-free freezers do this on purpose by circulating air so ice does not form on the inside of the freezer. It is not on the chart but the opposite of sublimation is deposition, which is when gas turns into a solid. This is how frost forms on a cold morning; water vapor turns to ice when it hits the cold grass or your car windshield.

Do you want to see some cool experiments with dry ice? Check out these links. Dry ice is fun. https://www.youtube.com/watch?v=kLO5SJ2uxEE

https://www.youtube.com/watch?v=pP_lZaOchE0

The changing states of water

Let's follow H$_2$0 or water through these phases of matter. Starting on a hot humid day with a lot of water vapor in the air, which is a gas, we get a big glass of ice water and sit it on the table. A few minutes later we notice water on the *outside* of the glass and dripping on the table! Where did it come from, the glass did not leak, it came from the air; the water vapor molecules in the air touched the cold glass and slowed down their movement, changing it into a liquid. This is **condensation** and what you might call "dew" in the morning. https://www.youtube.com/watch?v=bymT5AcV-C4 Now empty the glass and put it in the freezer you will now discover that the condensation on the glass is now solid ice, it has **frozen** just like frost on a cold morning on a car windshield. Put the glass back on the table and watch the ice **melting** back into liquid. Wait a while and the water is gone, it has **evaporated** back into a gas. Now, heat some water on the stove until it bubbles and make some tea. That of course is **boiling**. If you really want some fun, get some dry ice and put it in some water. The water will bubble as the dry ice **sublimates** into carbon dioxide gas. It also sinks in the water because it is denser than water. Throw in some dish soap and watch the soap bubbles overflow. Dry ice is fun, but do not touch it, your skin will freeze!

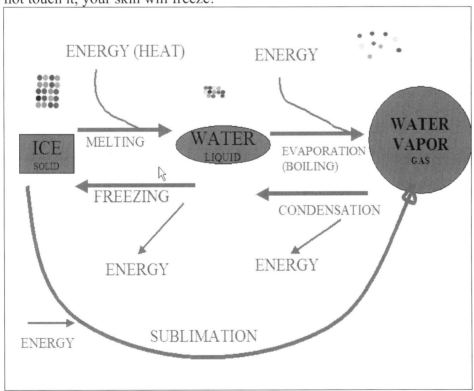

As you heat up ice the temperature *does not always change* – some of the heat is used to *change of phase* from solid to liquid or liquid to gas – this is why you cannot have a warm ice cube.

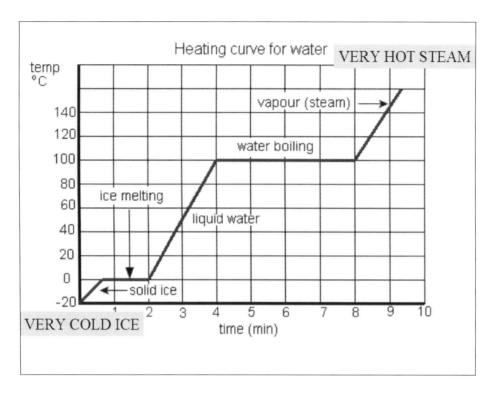

The water cycle

What we just described happens on earth all the time. Water goes into the sky by **evaporation** of water or **sublimation** of snow. The water vapor sticks to dust by **condensation** and **freezes** into ice crystals to form clouds. When the drops get heavy they fall as rain, or snow, where the process is repeated. Good thing we live on earth and not Mercury, because Mercury is too hot to ever have liquid water, which we cannot live without.

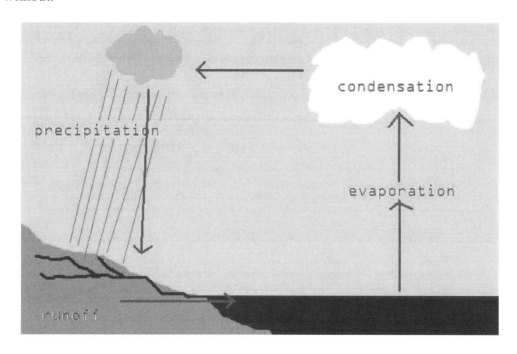

Other stuff can be in all 3 phases also

Ok, water can be a solid, liquid or a gas but what about metal? Sure it can, metals can be melted into liquid and even boiled into a gas at very high temperatures. It can then condense back into a liquid and freeze back into a solid. This is what industry does to make metal products from metal ore. Even air can be cooled enough to turn into a liquid or even a solid. Liquid Nitrogen is a good example; it turns into a liquid at –198 deg C (-321 deg. F) that is cold. A block of frozen nitrogen has to get down to –210 deg. C (-346 deg. F). Check out some experiments with liquid nitrogen at https://www.youtube.com/watch?v=KlAl5RHAAns, or https://www.youtube.com/watch?v=GYamb7WWyT4, or https://www.youtube.com/watch?v=ZgTTUuJZAFs.

The Mystery of Latent heat

Now for something really weird. The heat that comes out or goes in when water changes phase is hard to notice, it is hidden. **This is what latent heat means, hidden heat**. If you add heat to an ice cube it does not get hotter! It melts instead. **An ice cube is 0 deg C and when you add heat, the water it changes into is still 0 deg. C!** The heat was not used to make it hotter it was used to change the phase of matter. This is why you cannot have a hot ice cube. Even stranger is the fact that when you change water into ice, heat is removed and released into the air around it, and it actually increases the temperature of the air! This is why ice forms in a freezer, because the freezer actually removes the heat from in the fridge and out the back. This is why hot air comes out from behind your refrigerator, heat that used to be in the water.

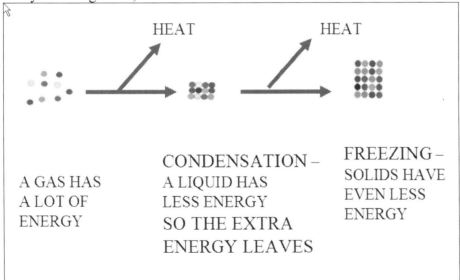

The heat leaving is the latent heat in this example.

One of the worse things that can happen to an orange tree plantation is frost. If the temperature gets below a certain point the oranges can freeze and the crop is ruined. This is expensive for the farmer, so they have some tricks that can help. They are using latent heat to warm the orange tree grove. It is confusing how this works but what they do is put out smudge pots, which are little fires that make a lot of smoke and dust. The fire and smoke do not heat up the air very much, but the water vapor condensing on the smoke dust does. As the gas turns into liquid on the dust, heat energy is released and the air is warmed hopefully enough to save the oranges. Very clever people, these farmers.

   Another time you may have noticed latent heat is in some types of instant hand warmers. These are hard to find now, but they contain of a special liquid in a bag with a metal disk. When you snap the disk the liquid instantly freezes into a solid and releases a lot of heat. The liquid inside was a material called Sodium Acetate. When you put the bag in boiling water, the sodium acetate melts (storing the latent heat) and when you make it freeze that heat is released. Some people call sodium acetate "hot ice" and it is very fun to play with. Check out a sodium acetate hand warmer at https://www.youtube.com/watch?v=PoXxDagbxCw or https://www.youtube.com/watch?v=4wv8Mi-Jl3E.

This is what happens when liquid sodium acetate is poured to make a castle, this is often called hot ice. https://www.youtube.com/watch?v=aC-KOYQsIvU

Here are directions to make your own sodium acetate,
http://chemistry.about.com/od/homeexperiments/a/make-hot-ice-sodium-acetate.htm

A cool thing you can try

Make snow - boil some water (it does have to be boiling) and take a cup of it outside on a *super* cold day. The temperature should be below –22 deg F though. Just throw the boiling water into the super cold air and you get snow! Make sure you do not throw the boiling water toward anyone or into the wind (it will go where you are) it can cause burns. I even made a small snowmaker machine with a power washer one winter, which is kind of how they make snow at ski resorts.

https://www.youtube.com/watch?v=p_f1PHKGQdg
https://www.youtube.com/watch?v=mlpYD9tdC8I

A day in the life of Earwig Hickson III

I learned today that my friend works at a welding shop where they use liquid nitrogen for something I really do not care about. What I care about is that he gave me some in a funny bottle with a very loose lid called a Dewar bottle. I guess that is so the escaping nitrogen gas will not build up pressure and pop the bottle. I wanted to observe the properties of solids at very low temperature. Wow, they become brittle.

I ran home to "study" my new toy, but what to do? I knew the stuff was bloody cold, like -320 deg. F, but what could I put in it. Looking around the house I found the perfect thing, my sister's tennis balls. I put them all in and waited while thick white mist boiled up around them. I took them out one by one with my mom's hot dog tongs and dropped them on the floor. Each one was really flattened and shattered into a million pieces when they hit the floor, a good start. Then I saw a vase of Roses my Dad had evidently brought home for my Mom, and since he was taking a nap and she had not returned home from work yet, I put them in the liquid nitrogen. Each shattered when I banged them against the table. I put hot dogs in, a balloon, ping-pong balls, a banana that made a great hammer, and a marshmallow chick. I was having a blast, all the food from the fridge went in, and then smashed. I was having so much fun I almost forgot to clean up the mess before my Mom got home, but managed to just in time. No damage done.

My Mom seemed in a good mood when she came in but suddenly she got mad at my Dad for some reason. The last words I heard as I ran out of the house were, "You forgot our Anniversary! Not even some flowers!" I decided to stay away from my Dad for quite a while. Dad ended up in the corner on *my* chair.

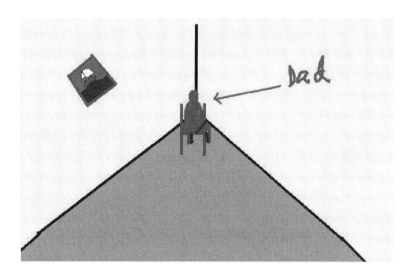

Review of terms – Quizlet:
https://Quizlet.com/124259265/chapter-11-volume-one-changing-states-of-matter-flash-cards/?new

Fun things to Google

Hot ice
Dry ice bubbles
Liquid nitrogen
Phet simulations – states of matter
Goldilocks zone for finding life on other planets
Turning boiling water into snow

Links

Building a fire on a bald mans head –
https://www.youtube.com/watch?v=smknECV63uI
https://www.youtube.com/watch?v=KvkLC-dSZYA

How to make an instant cloud in a bottle
https://www.youtube.com/watch?v=ft7q6efRIic

Snow making with a power washer -
https://www.youtube.com/watch?v=mlpYD9tdC8I

Making snow with boiling water – warning turn down the sound –
https://www.youtube.com/watch?v=p_f1PHKGQdg

Hot ice https://www.youtube.com/watch?v=oui2FYD00T4
https://www.youtube.com/watch?v=aC-KOYQsIvU

Instant hand warmer – when a liquid freezes!
https://www.youtube.com/watch?v=PoXxDagbxCw

Instant condensation – making a cloud in a bottle:
https://www.youtube.com/watch?v=OCmhvC27bto

Chapter 12
Using properties

Movie of the power point for this chapter:
https://www.youtube.com/watch?v=Ty5T3lv_46s

The wonders of science

A bird called the Arctic Tern migrates half way around the world and back again for a total of about 44,000 miles. Not only that but it finds the same nesting grounds each time. In fact many species of birds migrate each season and find the exact same nesting site they used in previous years. Salmon migrate from the ocean to the exact same freshwater stream where they were born. Monarch butterflies migrate from all over the United States and end up in central Mexico and it is not even the same butterflies, it is their offspring that arrive in Mexico, it can take five generations of butterflies to complete ONE migration. How weird is all this? I think really weird because I can't find anything without a GPS and some bird, butterfly or fish can migrate thousands of miles and not get lost. How do they do this? How to they navigate? How did it evolve? Well the answer is I do not know, and neither does anyone else (YET!). Do they use the Earth's magnetic field? The direction of the sun? Smell? Visual clues? Is it purely genetics? Maybe it is a combination of two or more?

There are many hypotheses trying to answer the question of animal migration and some discoveries have been made but we still do not understand it. We will one day, but not today. Scientists studying Salmon probably have the best grasp on the problem and have narrowed the answer down to two things, Magnetism and smell. Salmon evidently act differently around magnetic fields, so it is reasonable to assume, they may be able to detect the Earth's magnetic field, but they are not a compass, they *remember* what the magnetic field felt like in their brain when they were born. The property of the Earth's magnetic field is slightly different on different parts of the planet and the Salmon may be able to detect this. Wow. Not only that, but apparently the salmon also remember what chemicals were in the creek they were born in by a good sense of smell. Truly amazing! I suspect the answer (what ever it is) will be the same for birds, lobsters, elk, and the other animals that migrate.

Science is a never-ending process, with every discovery comes many more questions.

Monarch butterfly migration, https://en.wikipedia.org/wiki/Monarch_butterfly_migration

The Arctic Tern, https://en.wikipedia.org/wiki/Arctic_tern

How do birds navigate, http://education.nationalgeographic.org/media/how-do-birds-navigate/

Properties are used to identify things

Physical properties are used to identify things. Maybe you stepped in something and it stuck to your shoe. It might be a melted chocolate bar, but it might not. How can you tell?

Physical properties are the solution. Does it have a unique or recognizable odor? Is it magnetic? What is the density? What is the color? Is it sticky, liquidy, or chunky? All these physical properties are hints to the identity of the mysterious substance. Of course in science we never taste anything or sniff closely, bad things can happen.

Chemical properties can also be used to identify stuff. Is the stuff on your shoe flammable? Does it rust? Does it change the color of pH paper? These are useful hints to help identify an unknown substance too.

Some of the physical properties we use are color (green apples vs. ripe apples) odor (fresh fish vs. spoiled fish), density (remember Archimedes and the gold crown), magnetism (only Iron, Cobalt, and Nickel are magnetic, nothing else is). We can also see how well things conduct electricity (aluminum is fair, gold and copper are good conductors). You use physical properties all the time to identify things. You can tell the difference between a rabbit and a grizzly bear can't you?

The Chemical properties we use are the fun things to test. Flammability means the ability to catch on fire. Some things burn easily, others do not. Carbon dioxide puts fires out while oxygen gas makes them burn hotter. Some materials, like Iron, rust in the presence of oxygen, others like Gold do not, this is why jewelry is made of gold, not iron. Some chemicals change color in the presence of certain other materials. Iodine makes the starch in bread or paper turn purple; pH paper changes color in an acid or base. Baking soda bubbles when mixed with vinegar.

IDENTIFYING UNKNOWNS

- **PHYSICAL PROPERTIES**
- COLOR
- DENSITY
- MAGNETISM
- ELECTRICAL PROPERTIES
- OBSERVATIONS

- **CHEMICAL PROPERTIES**
- FLAMMABILITY
- CHEMICAL TESTS
- INDICATORS
- REACTIVITY

Using properties to separate mixtures

Remember chromatography? That was when we used the capillary motion of water to separate ink. How about distillation? That is one way to remove fresh water from polluted or salty water using boiling and condensation. Since water boils at 100 deg. C and most of the things in polluted water do not, they are left behind. The pure water is then condensed back into liquid. This could be used to purify polluted water for drinking.

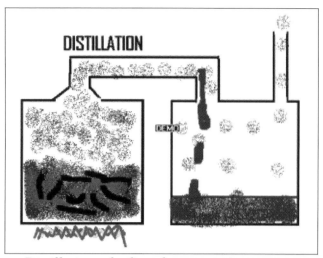

Distillation – boiling dirty water into water vapor and condensing into pure water

USING PROPERTIES TO SEPARATE MIXTURES

- MAGNETISM (SEPARATES IRON)
- FILTERS (FISH AQUARIUM, DRINKING WATER)
- EVAPORATION (SALT WATER)
- DISTILLATION (BOILING AND CONDENSING)
- CHROMATOGRAPHY

A day in the life of Earwig Hickson III

Today I decided to make my own water purification system. I thought it would be cool to make my own drinking water. To make sure it worked under the most extreme conditions I ran down to the local sewage treatment plant and stole a few gallons of nice raw sewage right from the pipe. I did not think anyone would mind but the Forman called the police on me, so I ran.

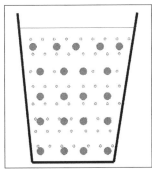

When I got home and opened the jug it was not pleasant. Nothing about the color or smell made me want to taste it so off to the lab I went. First I poured the mess through a fine screen mesh from our porch door, it took the chunks out but it was still foul. I then poured it through a thick sweater I borrowed from my sister's closet. It actually helped a lot, almost all the floaters I could see were removed. I returned the sweater to the closet.

As I stared at my semi-transparent solution, it occurred to me that I could borrow the charcoal filled cartridge from my Dad's fish aquarium filter. Wow what an improvement the solution was nearly clear, but the smell was still a little odd. I returned the filter before my dad found out. The fish aquarium water was a little cloudy for a while and the fish were not happy.

To test my progress I filled my sister's water bottle with it and put it back in the fridge. She did not like it at all, it attracted flies and after smelling it she dumped in down the drain! My project was gone. What a waste. Next time perhaps I should add a squeeze of lemon.

My next improvement was to boil the solution and see if that could knock the odor back a bit. It would at least sterilize it. Success, quite an improvement! I could hardly tell it was sewage only a few days before. I gave some to the cat and she drank it, then again she drinks from the toilet. I needed a better test subject. I tried adding it to the coffee maker, made some Kool-aid drink and serving it with ice at dinner. From my side everything looked fine, but my dad made a call to the water company to find out why the water tasted odd. They all spent the majority of the next day in the bathroom, I am glad I drank milk. I guess that was a fail after all.

Then it occurred to me that I could boil the solution and trap the water vapor through the process of condensation. I made my own distillation machine. The steam cooled off when going through a long pipe, clear water dripped out. It looked clear, it had no smell, and it actually looked good.

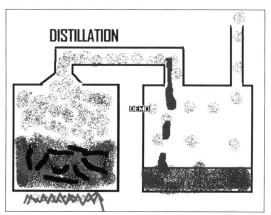

The final test was at dinner. I served everyone my self-purified water. Everyone thought it was good, but did ask me a lot, why I kept asking how he or she liked it. I felt so proud of myself; I successfully turned sewage water into pure water. In my excitement I told them what I had done, expecting praise for my ingenuity! I did not get praise, I got grounded again and had to unhook our sewage line from the water line coming into the kitchen sink. As a bonus though, I never had to help with dinner again.

Now that I think about it, I am not sure I trust my Dad. He always looks at me funny when I drink my orange juice in the morning, and leaves the room with a silly smirk.

I have to go to the bathroom again.

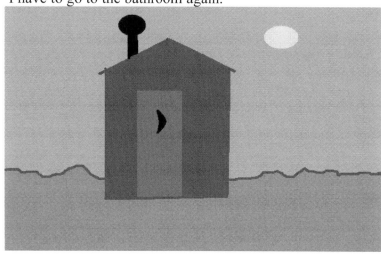

Review of terms – Quizlet:
https://Quizlet.com/124260068/chapter-12-volume-one-using-properties-flash-cards/?new

Fun things to Google

Paper chromatography
Distillation
How to pan for gold
Archimedes and the gold crown
What are indicators?

Links

How to identify counterfeit money using properties of matter.
http://ed.ted.com/lessons/how-to-spot-a-counterfeit-bill-tien-nguyen

Where do we get our water? A Ted-ed film. http://ed.ted.com/lessons/where-we-get-our-fresh-water-christiana-z-peppard

What is chromatography? How can you do it? A Ted-ed film.
http://ed.ted.com/featured/4bURoBhk

How does the international space station purify water?
https://www.youtube.com/watch?v=KuPMR_vMNR0
https://www.youtube.com/watch?v=BCjH3k5gODI

How is sewage cleaned?
https://www.youtube.com/watch?v=gxgpK1EUZns

Where does your drinking water come from?
https://www.youtube.com/watch?v=9z14l51ISwg

A simple test to identify counterfeit money using tincture of iodine:
https://www.youtube.com/watch?v=JEO-zFz9fs0

Unit 4
Energy

Chapter 13
Energy forms and types

Movie of the power point for this chapter:
https://www.youtube.com/watch?v=U79ZkflLsiQ

The wonders of science

On the morning of June 30, 1908, there was an explosion, a big explosion, a very big explosion. One thousand times bigger than the atomic bomb dropped on Hiroshima, Japan to help end WWII. This giant explosion is known as the Tunguska Event, and it happened in a sparsely populated area in Siberia (Russia). It knocked down trees in an area of about 800 square miles, every tree within 16 miles, creating a circle of devastation 32 miles in diameter. What could possibly create such an explosion, nearly 40 years before nuclear bombs were invented? Could it be an alien space ship or maybe an underground natural gas explosion? The probable answer is a large asteroid about 200 – 600 feet in diameter (600 feet is two football fields). The asteroid evidently exploded somewhere between 3 to 6 miles above the Earth's surface knocking down 80 million trees.

There are always rocks hitting Earth from space, most of them small. We see these as shooting stars (about the size of a grain of sand) but sometimes a bigger rock will hit resulting in a small explosion and even a crater. Sixty-five million years ago an asteroid roughly six miles across hit our planet and wiped out nearly all life (including the dinosaurs) because it changed the climate from all the dust it released into the atmosphere.

The energy of the Tunguska Event explosion is estimated to be from 10 to 30 megatons of dynamite. That is a lot of energy.

The Tunguska Event, https://en.wikipedia.org/wiki/Tunguska_event

There are a lot of forms of energy

Energy causes changes to matter. You can usually see this, the matter changing at least, but not the energy. Energy going into matter *causes* something to happen, lights to turn on, temperature to go up, objects change their motion, things change in color, something always happens. Energy can be removed from matter to cause change also, like water turned into ice. You simply put water in the freezer; the freezer removes energy from the water, slowing down the molecules until they stop sliding around. The result is a solid with less energy than the liquid had originally. You can see all this because you can see matter, the energy causes all this but you cannot see it, you might be able to feel it or hear it but you cannot see, it is invisible.

It is kind of like the deep base in a song, it shakes your bones and teeth or even a whole car, but you cannot see the sound energy, only the matter being affected by it. You can't see gravity but you can see things falling. This is because energy has no mass or volume; you cannot put it in a jar and get out it later.

Forms of energy

Mechanical energy

Mechanical energy is the energy in a *moving* object. A moving car has a lot of mechanical energy, so avoid them running into you. Anything moving has mechanical energy. Mechanical energy can hurt. Running goats have mechanical energy. Falling rocks have mechanical energy. It is moving objects. Mechanical energy is easy to recognize, we are familiar with it. It can whack us, and that can hurt.

Sound (acoustic) energy

Sound comes from a *vibrating* object (*the <u>whole object</u> not the individual molecules*), like a guitar string, tuning fork, vocal chords, a bell, a crashing car, or a cell phone ringing in the middle of class.

When an object vibrates back and forth it *hits* air molecules and that forms sound waves of energy. When these waves hit your eardrum they vibrate that too and your brain interprets that as a sound. A sound is energy so sometimes a sound wave can hit something and cause it to shake so much that it causes damage, like a sonic boom from a faster than sound jet plane (which can break windows). Someone may ask if a tree that falls in the woods makes a sound, I don't know, well maybe I do, it does make a sound wave, because it vibrated when it crashed. It made energy, and that makes matter vibrate, and that is sound, it made a sound.

A singer with a powerful voice can sing a loud note (at just the right frequency) and shatter a glass, this was proven on Mythbusters (time to Google). The energy of the singer's sound wave made the glass vibrate so much that it broke, amazing.

Play with energy waves at Phet - https://phet.colorado.edu/en/simulation/wave-on-a-string

Experiment with sound waves with this another cool simulation from Phet - https://phet.colorado.edu/en/simulation/sound

Another fantastic simulation by Phet where you can play with sound, light and water waves - https://phet.colorado.edu/en/simulation/wave-interference

And yet another Phet game using radio waves - https://phet.colorado.edu/en/simulation/radio-waves

An interesting thing about sound waves is that since they move so slowly through air (at *only* 765 miles per hour), sometimes they get squished and cause a change in pitch. This is called the Doppler Effect. https://www.youtube.com/watch?v=imoxDcn2Sgo

When a car (or better yet a police car with a siren) moves quickly by you, you may notice a change from high pitch (coming towards you) to a low pitch (going away from you). A mosquito flying around your head does the same thing. The change in pitch (or frequency) is from the car (or mosquito) flying the same direction as the sound it produces and causing the waves to get smooshed together, resulting in a higher frequency. When it moves away the sound waves are stretched apart resulting in a lower frequency. https://www.youtube.com/watch?v=eo_owZ2UK7E or https://www.youtube.com/watch?v=h4OnBYrbCjY

Sometimes when you are watching a baseball game from far away, you may notice that you see the batter hit the ball before you hear it. This is because light travels much faster (186,000 miles per second) than sound (0.2 miles per second). Maybe you noticed that you see lightning before you hear thunder (they are produced at the same time) for the same reason. This is why you can count the seconds between lightning and thunder and estimate how far away the lightning strike was. Five seconds means about one mile. https://www.youtube.com/watch?v=Sp9bKDHRfsM Here is a nuclear bomb going off from about seven miles away – you see it way before you can hear it. https://www.youtube.com/watch?v=Mn7PeI2UyEM

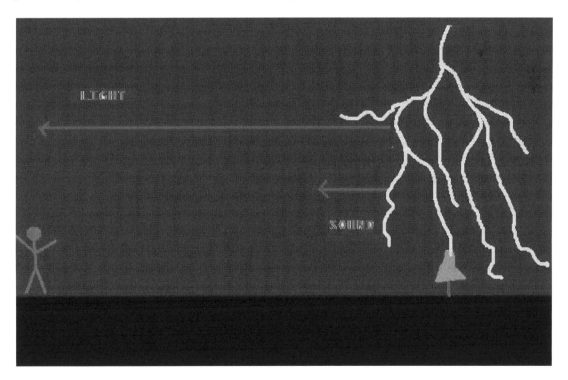

Chemical Energy

This is the energy trapped in chemical bonds and the hardest to understand. It is the energy in food, oil, coal, gasoline and wood for the fire. All of this energy originally came from the sun. It was captured on earth by plants via photosynthesis. A plant used the sunlight to combine carbon dioxide and water into a bigger molecule called sugar (glucose); this sugar is food for us and other animals. Chemical energy is stored sunlight (but it is not light anymore). When we eat the sugar the opposite happens, this is called respiration. A short explanation by Ted-ed,
https://www.youtube.com/watch?v=eo5XndJaz-Y

Weird as it sounds, when you eat a piece of yummy bacon, it is not the atoms and molecules (mass) that give you energy; it is the chemical energy of the chemical bonds holding the atoms together. *Extra* energy is used by your body to make fat, which is stored energy (your body uses the energy to put the fat molecules together which makes you gain weight). I have been told I have a lot of stored chemical energy; ask your teachers if they do too.

Photosynthesis
Carbon dioxide + water +sunlight energy → sugar + oxygen

When we eat food our bodies do the opposite, which is called *respiration*.

Respiration
Sugar + oxygen → carbon dioxide + water + energy

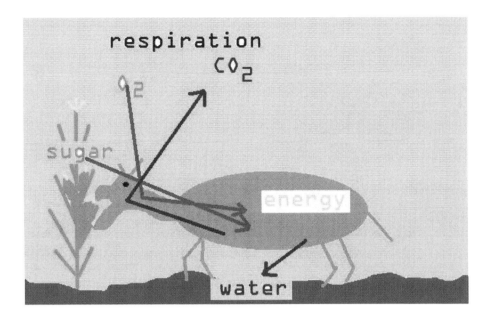

Thermal energy

Thermal energy is the *movement or shaking of individual molecules* in an object (not the whole object). *It is not heat.* Heat is the *movement* of thermal energy (more on that later). All objects have thermal energy, even very cold objects, just not as much.

Every molecule on earth (and even the universe) has some energy, this makes them shake or move. As the temperature gets higher, they move faster, storing the energy. If the temperature gets lower they slow down. Eventually they could stop moving when ALL the thermal energy is gone. This temperature is called absolute zero; it is about negative 473 deg. F. No one knows what happens at absolute zero; scientists have only managed to get down to a fraction of a degree above it, using really advanced thermoses.

I am going to mention now, that there is no such thing as cold, only *thermal energy*. Instead of saying something is cold, you should say it has *very little thermal energy* (or *lacks heat* if you want). When you get yelled at by the boss at home, for letting the cold into your house, when you left the door open on a winter day, correct them by saying, "I am not letting the cold in (there is no such thing) I am letting the heat out". When you walk into a freezing cold room, you should say, "There is a lack of thermal energy in here." People will be impressed, but they might wonder about you.

Phet gas and heat properties - https://phet.colorado.edu/en/simulation/gas-properties

Electromagnetic Energy (light)

Electromagnetic energy is what you think of as *light*. Humans can only see visible light, but there are other kinds of light we cannot see. The electromagnetic spectrum is all the kinds of *light*. They include microwaves, infrared light, radio waves, gamma rays, X-rays, ultraviolet light and of course visible light.

Some of these forms of light can cause damage to living things (it is energy after all). X-rays can go through skin, which is why they are used to look at the bones inside your body (they can't go through bone). Gamma rays go through everything, and cause cell damage or cancer. Some forms of ultraviolet light cause sunburn and even skin cancer. Infrared light is heat, and is used to keep food warm in a restaurant. This is why food kept under those hot lights is hot; it is also a little mushy.

I find it interesting that some animals can see light we cannot. Insects can see ultraviolet light (invisible to humans); they see colors differently than we do. A yellow flower to us may be purple to a honeybee. A white female butterfly may look red to a white male butterfly, while the white male may look orange to the female. Strange. https://www.youtube.com/watch?v=N1TUDFCOwjY Cats and dogs have vision adapted to nighttime. There is evidence that they might see in the infrared range and probably the ultraviolet too, either way they see the world in a totally different way than we do. My cat hisses at nothing sometimes, maybe there is something there after all! Perhaps I should worry.

You may have a black light at home; this is a special light that makes ultraviolet light (which humans cannot see, thus it is black to us). Some pigments (colors), you put under the light, have the ability to change the ultraviolet light, into visible light (we can see), this is why they glow. Some companies put ultraviolet dyes in their products to trick us. Toothpaste and laundry detergent contains these dyes so your teeth and white socks look cleaner. Try them under a black light and you will see your teeth glow; at least they do if you brush them.

There is a small part of the electromagnetic spectrum that humans can see. This is called visible light, the colors of the rainbow. ROYGBIV, which stands for Red, Orange, Yellow, Green, Blue, Indigo, and Violet. When all these colors of light are mixed together we see WHITE light (sunlight). If you shine that white light through a prism, you see a rainbow (ROYGBIV). Weird. Here is a short video about the electromagnetic spectrum. https://www.youtube.com/watch?v=m4t7gTmBK3g

Nuclear energy

Nuclear energy scares some people, and rightly so. It is the energy that comes from the nucleus of an atom that has been split in half or slowly decays. In the wrong hands this can be bad. Nuclear energy can be used to make clean inexpensive electricity or a nuclear bomb. The waste is radioactive and hard to dispose of. Some natural elements are radioactive, such as uranium and Radon. This is because these atoms are so large that they fall apart by themselves and release a lot of energy in the form of radiation (which is bad). https://www.youtube.com/watch?v=S-Ld1L0TObQ

Nuclear energy has many good uses though. In a nuclear power plant, uranium is allowed to decay slowly (over 15 years or so) producing heat. The heat is used to boil water into steam and turn a generator (more on this later). The generator sends electricity to your house. The radiation is contained in an underground thick walled building and when working properly is safe. By the way the electricity is just like any other electricity, it is NOT radioactive. https://www.youtube.com/watch?v=d7LO8lL4Ai4

There have been accidents though. In the United States, a nuclear power plant called *Three Mile Island* had a malfunction and a partial melt down. Not much radiation escaped and the area is safe to live in today. In fact I currently live 8 miles away and work 2 miles away from it. I see it every morning on my way to work. In Russia a nuclear reactor in Chernobyl had a *total* meltdown. So bad in fact that no one can live in that area anymore, the whole city is abandoned, hundreds of square miles where no people can ever live again! The area is not dead though. It is too radioactive for people but animals and plants live there; in fact it is one of the most wilderness areas on the planet, where nature can do what it wants without people messing with it. It is a fascinating ecosystem and one that many scientists are studying. This is a rather long interesting video about the wolves that live there now, https://www.youtube.com/watch?v=2Op2ZIg4JxE

Nuclear energy is used in medicine too. It is used to treat cancer in patients (radiation therapy) and a variety of things (hit the internet if you are curious). Radiation is also used to kill germs on food, thus making it stay fresh longer. This food is not radioactive by the way, and probably healthier that non-radiated food.

Another interesting bit about nuclear energy is that the Sun is a giant nuclear reactor. I mean ginormous. It has been burning for at least 5 billion years and we can expect it to burn for another 4 billion. There is a lot of energy in the nucleus of atoms and the sun has A LOT of atoms.

Did you know you are surrounded by radioactive things? This is called background radiation. The smoke detector in your house has a radioactive material in it. It is very little so it is safe. You should have a smoke detector in your house.

Other radioactive things in your house may not be so good. Some houses contain Radon gas, which can cause lung cancer. If you have a basement with a lot of cracks, it is worth buying a Radon detector device. You can get them at a home improvement store and it is worth it. Radon is BAD.

Electrical energy

Electricity is *moving electrons*. They start at a power plant and follow wires to your house. When the electrons hit things like the inside of your T.V. they change into light, sound and heat. Our society cannot survive without electricity, and I wouldn't be able to watch any Philadelphia Eagles football games.

Electricity can be fun and is always useful. There are two things to know about electricity. Voltage is how much and amperage is how fast it moves. Amperage can kill you, voltage will not. Low amperage but high voltage (like an electric fence) is not dangerous. High amperage but low voltage is dangerous (like the outlet in your house). Grab an electric fence on a farm if you feel the need to, but *never* put your finger in a light socket or wall plug, you will get fried. One of the funniest things I ever saw was a hunting dog peeing on an electric fence. Do not try that at home!

Static electricity game from Phet - https://phet.colorado.edu/en/simulation/john-travoltage

Another static electricity game from Phet - https://phet.colorado.edu/en/simulation/balloons-and-static-electricity

Make an electric circuit from Phet - https://phet.colorado.edu/en/simulation/circuit-construction-kit-dc

Magnets and magnetic energy

Magnetism can work at a distance. This is why the earth (which is a big magnet) can be used to make a compass point north.

How magnets work by Phet - https://phet.colorado.edu/en/simulation/magnets-and-electromagnets

Gravitational energy

This is energy caused by the earth's gravity. It makes things fall. But you know all about gravity, don't you?

Elastic energy

This is energy stored in a spring or rubber band. It kind of makes sense.

Review

	NAME	DISCRIPTION
• 1	M ECHANICAL	MOTION
• 2	S OUND	VIBRATIONS
• 3	C HEMICAL	CHEMICAL BONDS
• 4	T HERMAL	PARTICLE MOVEMENT
• 5	E LECTROMAGNETIC	LIGHT
• 6	N UCLEAR	ATOM
• 7	E LECTRICAL	ELECTRONS
8	GRAVITATIONAL	GRAVITY
9	ELASTIC	A SPRING

FORMS OF ENERGY

The two <u>types</u> of energy

There are a lot of forms of energy but only two types, energy that is moving, and energy that could move at any time. Kinetic energy is the energy of *motion* and Potential energy is the energy of *position*.

Potential energy

Let us imagine that I have placed a large rock hanging from a string on the ceiling of my classroom right above a kid's head. That rock hanging there is harmless and not hurting anyone. It is stationary but since it *could* fall or has the *potential* to fall, it has potential energy. Or more specifically *potential gravitational* energy, since gravity is what could make it fall. But since it only has potential energy it is harmless. Potential energy is always *harmless* as long as it *stays* potential energy.

Potential energy is often times called the *energy of position*, or stored energy. The rock hanging from the ceiling is a threat, if it was on the floor, no threat. A rock on the floor has nowhere to fall, the one on the ceiling does.

There are other ways energy can be stored. A lake far up in the mountains wants to flow down hill to the ocean. It has a lot of *potential gravitational* energy. If you do not believe me, imagine a large lake behind a dam, if the dam was to break, what would happen below it? Major destruction I assure you. This has happened, check out the "Great Flood" in Johnstown PA. All lakes have a lot of potential gravitational energy.

Batteries have potential energy too. The energy inside them is actually chemical energy, but they have the potential to produce electrical energy. They are said to have *potential chemical* energy, they are not filled with electricity, just chemicals. When you stretch a rubber band to shoot it across the room, just before you let it go it has *potential elastic* energy because it could "fire" at any moment. Same with a set mousetrap.

Some of the forms of energy that can be potential (because the energy is there) are:
Potential-gravitational - since an object can fall or go downhill at any moment
Potential-elastic – a stretched spring or rubber band, ready to snap.
Potential-nuclear – An unstable atom can decay at any time.
Potential chemical – food has potential energy, so does coal or gas before you burn it.

Some forms of energy that cannot be potential (because the energy is not there) are:
Acoustic (sound) – you really cannot have sound ready to go, you can have a bell but there is no sound in it, energy must be added to the bell (by shaking) then the bell converts it into sound.
Electromagnetic (light) – The same idea, you just can't fill up a jar with a light beam and expect it to still be in there the next day.

Kinetic Energy

Kinetic energy is the *energy of motion* (not to be confused with mechanical energy). Imagine if the rock that was hanging by a string with only potential energy, was allowed to fall when the string broke, or was cut. Now that it is moving, it has kinetic energy and could hurt anyone below it. Kinetic energy does *work* (it changes something). Moving objects, like cars have *kinetic mechanical* energy. Electricity flowing out of a battery is *kinetic electrical energy*. A river flowing out of a mountain lake has *kinetic mechanical energy*.

Forms of energy that are kinetic

Kinetic mechanical – Any moving object like a car, river, rolling boulder, falling rain, hail, a swinging baseball bat, or the ball after it is hit.
Kinetic acoustic – Anything you can hear, a sound because the sound wave is moving across the room.
Kinetic thermal – Heat warming you from a fire or burning your hand when you touch a hot stove. The thermal energy is moving into your hand.
Kinetic electromagnetic – Someone shining a flashlight into your eyes, the light is moving.
Kinetic nuclear – No need to explain – poof no more eyebrows, or hair, or skin, or bone, energy moves away as heat and light.
Kinetic electrical – Moving electricity like an electric wire or grabbing an electric fence.

Kinetic gravitational – A falling rock or falling off a cliff.
Kinetic elastic – When a mousetrap snaps.

An absolute awesome simulation about potential and kinetic energy is from Phet –
Energy Skate Park - https://phet.colorado.edu/en/simulation/energy-skate-park
https://phet.colorado.edu/en/simulation/energy-skate-park-basics

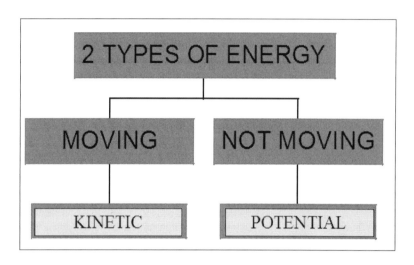

The deer

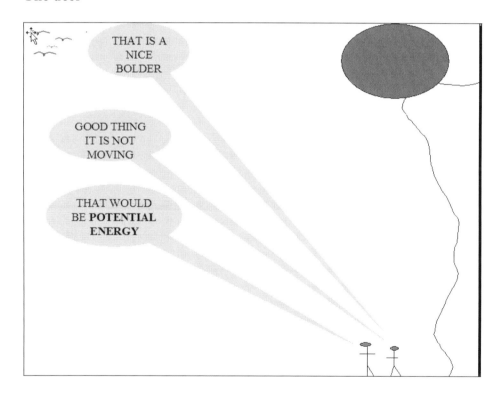

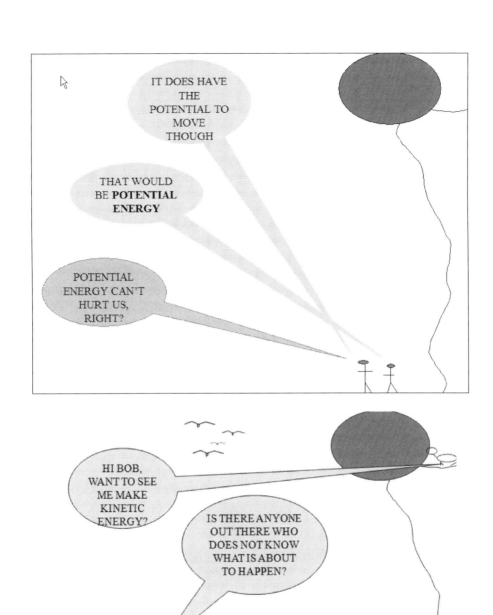

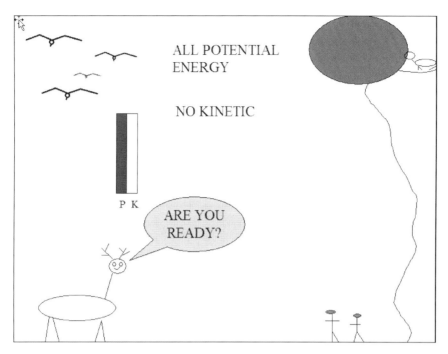

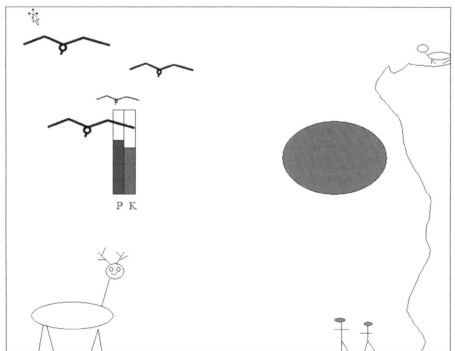

142

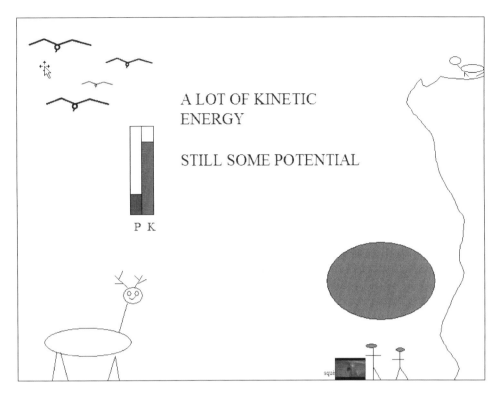

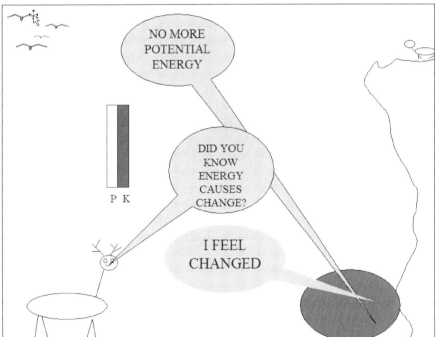

When the rock lands and stops, it no longer has any energy. The energy is somewhere but not in the rock, it changed into sound and heat energy and is no longer noticed.

A day in the life of Earwig Hickson III

Today my family went to the museum but since I was still grounded from the drinking water experiment and they did not want me *getting any ideas*, I had to stay home and do chores. I was to clean my room. I was starting my task when I accidentally knocked a book off my desk and it landed on the edge of a dirty spoon that flipped up into the air and landed on my chair, where it hit a ball that rolled off and bounced across the room and just missed knocking over a glass full of old chocolate milk. At first I felt lucky the glass did not spill and make a bigger mess, but then I thought, perhaps with a little adjustment I could hit it next time. I replaced everything exactly as it was and adjusted the ball a bit. Bull's-eye! I got the glass, I was on fire, and I knew what to do next! I was going to build a Rube Goldberg Machine. It would be cool.

I knew a *Rube Goldberg Machine* consisted of many parts working together in a system. It needed to have many energy conversions always changing potential energy into Kinetic energy and back again. I was going to use all the forms of energy I could, then film it and put it on YouTube where it would go viral. I would be famous!

I started with a bowling ball hanging from a ceiling hook, it would land in my toy truck on a ramp and start a giant chain reaction of toys, dominos, ramps, levers, balls, hanging buckets of water, electrical switches, mousetraps, a rat trap I found with a petrified rat in it, and everything in the house I could find. It was wonderful, it began in my room with the bowling ball, left my room when a toy car rode out the door on a race track, went down the hall and into the bathroom where it filled the bathtub with water, down the steps when another bowling ball was hit by a sailboat. At the bottom of the steps a compressed spring shot a crochet ball into a light switch, which turned on a blender, which vibrated and fell off the counter onto a bucket tied to a string, which

144

pulled a barbell off the table. Eventually the machine went out into the living room energy changing types and forms the whole way. I used a lot of potential energy by hanging things from the ceiling with nails from the garage. It turned into kinetic energy when the things fell. I used elastic energy by compressing springs, gravitational energy by letting things fall, electrical energy with my erector set robot, acoustic energy with a bell and there was mechanical energy everywhere. It made thermal energy when a match struck and made a small fire in the basement, chemical energy when it lit a firecracker making more sound and some electromagnetic light. I didn't use nuclear energy though; the lady on the phone wouldn't let me have any and sounded concerned with my reason why.

My room was clean with everything in its place; only "its" place was a part of my wonderful machine. When I triggered it all kinds of things happened, not all planned, the house was a wreck, my room was messy again, there was a hole through the front window but that was not my fault, the cat did it when the bowling ball bouncing down the steps landed on his tail. The nails in the walls and ceiling I figured could be used to hang house decorations. The crack in the T.V. was harder to justify, the water damage from the waterfall, impossible. To this day, I regret using the electric fence and the cow that followed me home from the farm.

If my father were Albert Einstein he would have nurtured my creativity and surely would have praised my ingenuity. Maybe sent me to MIT as the youngest student in history! But Albert Einstein is not my father. It was a long time before I was left alone at home or allowed back into the house. I like my tent but it does still smell bad. Kitty helps keep me warm.

Review of terms – Quizlet:
https://Quizlet.com/124260627/chapter-13-volume-one-energy-forms-and-types-flash-cards/?new and
https://Quizlet.com/107241557/chapter-13-volume-one-energy-types-and-forms-part-2-flash-cards/

Fun things to Google

Ultraviolet vision in bees
Ultraviolet vision in cats and dogs
Night vision goggles (YouTube)
The Doppler Effect (YouTube)
The Doppler Effect the big bang theory (YouTube)
The gummy bear experiment (YouTube)
Mixing light
Rube Goldberg device – machines that are fun to watch and build (YouTube)
Three Mile Island accident
Chernobyl accident
Chernobyl life in the dead zone – it is amazing what lives in a nuclear wasteland
Playing with an electric fence (YouTube)
Johnstown PA flood
Watermelon drop
Sonic boom
SR-71
Mythbusters, can a human voice shatter a glass
Liquid nitrogen – cold stuff
Absolute zero
Radon in your home

Links

Kinetic and potential energy explained by Newton and Napoleon
https://www.youtube.com/watch?v=CK1y-jSiSoQ

The gummy bear experiment for the first time.
https://www.youtube.com/watch?v=RxC6I8JN6ec

Color vision – you know that if you mix all the colors of paint you get black but what about all the colors of light?
https://phet.colorado.edu/en/simulation/color-vision

Phet Energy skate park – this is a lot of fun and educational
https://phet.colorado.edu/en/simulation/energy-skate-park-basics

Bowling ball of death camera view
https://www.youtube.com/watch?v=49jIqpW1Oow

Phet – Gas properties – try it, you can get to absolute zero (0 deg Kelvin). Fun and educational.
https://phet.colorado.edu/en/simulation/gas-properties

How X-rays see through your skin. A short film by Ted-ed.
http://ed.ted.com/lessons/how-x-rays-see-through-your-skin-ge-wang

What is a sonic boom? A simple sound wave? http://ed.ted.com/lessons/what-causes-sonic-booms-katerina-kaouri

Light we can see and light we cannot. A film from Ted-ed. http://ed.ted.com/lessons/light-waves-visible-and-invisible-lucianne-walkowicz

Converting potential energy into kinetic energy and back again: https://www.youtube.com/watch?v=IqIjBW0aiTU

Chapter 14
Energy transitions

Movie of the power point for this chapter:
https://www.youtube.com/watch?v=-pOLTW8RRjM

The wonders of science

The water wheel was invented sometime around 400 BC. It could replace 50 laborers doing the job faster and cheaper. It was the first power source that could be used to run industry. In America, in the 1700's water wheels were responsible for the expansion of the early colonies. You see when people settle a new land, they have to supply themselves with food, and food means farms, and farms mean the grain must be processed into flour. In any system (like this food system) there is always a limiting factor that slows down the process. In this case it was grinding the grain. A person could do it for his or her own use but to make money and feed the local town a lot of grain had to be ground. Thus the water wheel saved the day. A water wheel is a giant wheel that is turned by running water; the wheel is attached to an axle that spins also. The axle turns a series of gears that put the energy where you want it. In this case the gears turned a big millstone, which could grind the grain quickly, cheaply, with few workers. This was called a gristmill. As people expanded into new areas towns sprang up around the gristmill. Now everyone could eat, the population could increase, and American expansion accelerated.

As towns became bigger new industries had to be built. Blacksmith shops, sawmills, furniture factories, paper mills, textile factories, and tanneries. Any machine that uses an electric motor today could use water wheel power 300 years ago. The watermill converted the energy of running water into useful work for society. Science saves the world (again).

The water wheel, https://en.wikipedia.org/wiki/Water_wheel
The watermill, https://en.wikipedia.org/wiki/Watermill

Energy transitions

A transition means a *change*. In the case of energy it means a change from one type or form of energy to a *new one*. Now in the case of energy (which is what we are talking about), it cannot be destroyed, it can go somewhere where you do not notice it, but it is somewhere. This is why we scientists say *energy is never lost, but it can change from one form to another.* We use this idea to change *useless* energy (like a river) to energy our society needs (like electricity). It is useful to review that almost all the energy we use (excluding nuclear) comes from the sun. The sun is our savior, without it, the earth would be *dead*, no rain, no wind, no wood, no food, no coal, no gas, no rivers, no sound, no light, no nothing. The planet would be a dead rock. The sun is great! We depend on it.

Now the easiest way to understand this is the conversion from potential energy (energy of position) to kinetic energy (the energy of motion). Rivers start in mountains and have a lot of *potential energy*; they flow down hill with *kinetic energy*. The water in the clouds has *potential energy* but rain, snow and hail have *kinetic energy*. Wood has potential energy but a fire has kinetic energy. Electricity can be turned into light and many other things. Gasoline can be changed into the motion of a car. Potential energy can easily be changed into kinetic energy. This is a good time to Google *Phet Skate Park*, it is fun. https://phet.colorado.edu/en/simulation/energy-skate-park

Energy is NEVER lost, it just changes into another form of energy, we may not notice. In our homes it changes into light, sound and heat, which is not useful at the moment and we think it goes away, but it does not, it just is not noticed. For example the electricity that goes into your T.V. makes it possible to watch a movie, but it turns into useless heat, light and sounds afterwards.

Electricity comes into your homes. Electricity is useless for our needs unless we can change it into something more useful. We do this with *appliances*. An appliance is any machine that changes one form of energy into another kind. An electric stove is an appliance because it changes electricity into heat. A television, radio, washing machine, light bulb, toaster, and even a blender are all appliances because they change electricity into a different kind of energy that can do a useful job for us.

The law of conservation of energy

The law of conservation of energy says that *energy cannot be created* (from nothing) *or destroyed* (into nothing) *but it can change forms*. What this means to us, is that useful energy (like electricity) changes into energy we cannot use (heat light and heat), which we think is gone but it is not, it is just changed. On the other hand useless energy (like wind) can be changed into something useful like electricity. This is what makes our society run. We need energy, specifically electricity.

Phet – energy changes - https://phet.colorado.edu/en/simulation/energy-forms-and-changes

The generator

To make electricity we use a machine called a *generator (also called an alternator)*, there are other ways but this is the most reliable. A generator is like a backwards motor, except instead of electricity turning the motor, the motor turns to make electricity. So the problem is how can we turn the generator? There are a lot ways to turn a generator; it starts with a propeller, if you can turn the propeller (or more properly a *turbine*), the generator turns. So how do you spin the turbine? There are a lot of ways to do this too. Water will turn the turbine or wind, but usually we use steam. Steam is easy to control and very efficient. So how do we make steam? By burning something like coal, oil, or natural gas. The point is it does not matter HOW you turn the generator, only that you do.

Phet – How generator works - https://phet.colorado.edu/en/simulation/generator

Phet energy changes - https://phet.colorado.edu/en/simulation/energy-forms-and-changes

Hydroelectric power and where the energy comes from and goes

In this story, it all starts with sunlight. The sun sends electromagnetic radiation (light) to the earth, which causes water to evaporate into the sky. This makes it gain altitude and condense into a cloud, which has a lot of potential energy, since it is so high in the sky.

By the way even though a cloud is light and fluffy, it is huge and contains thousands of gallons of water, it is heavy.

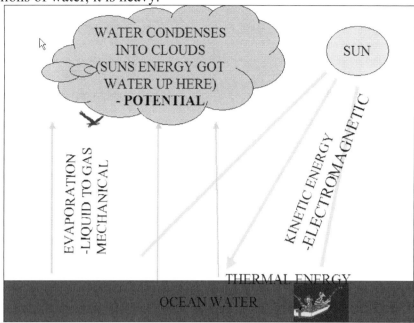

The cloud moves over land because of wind (also caused from the sun). The wind by the way can be used to spin a wind turbine also.

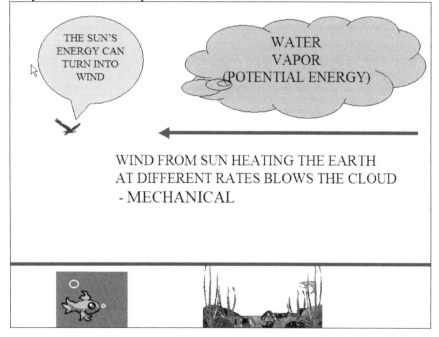

\

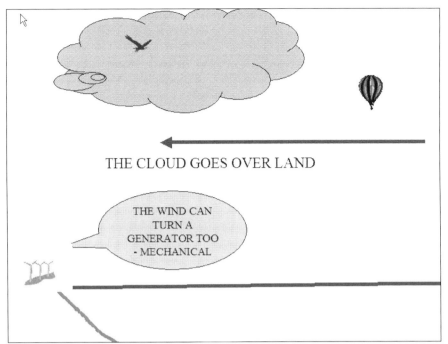

Eventually the water droplets in the clouds become too heavy to stay up there and raindrops fall.

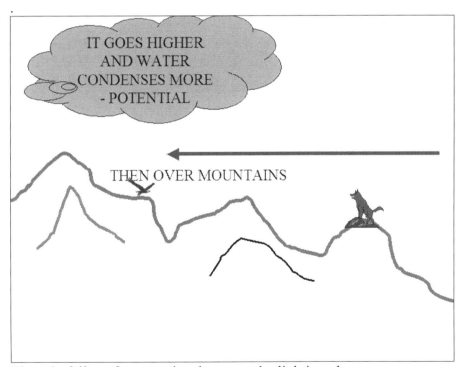

The rain falls to form creeks, there may be lighting also.

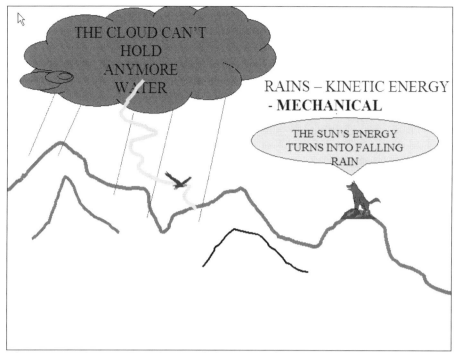

The streams have a lot of potential energy since they are high up in the mountains and flow downhill with kinetic energy.

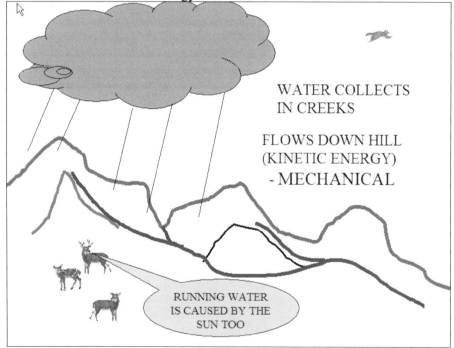

Eventually the river flows into a man made lake (one with a dam) and stops. The lake has a ton of potential energy since it is still uphill in the mountains.

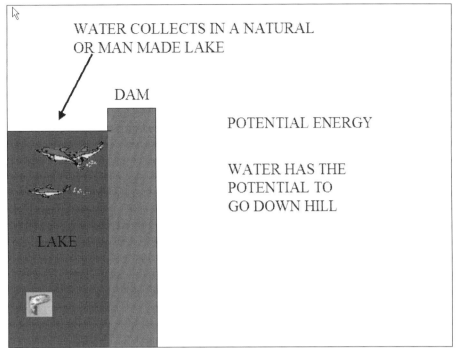

WATER COLLECTS IN A NATURAL
OR MAN MADE LAKE

DAM

POTENTIAL ENERGY

WATER HAS THE
POTENTIAL TO
GO DOWN HILL

LAKE

In the dam there is a tunnel near the bottom of the dam (where the pressure is highest) and the water squirts through. Inside the tunnel of fast water, is a turbine (propeller), which spins very fast.

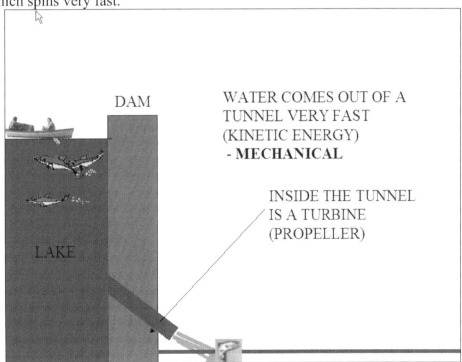

DAM

WATER COMES OUT OF A
TUNNEL VERY FAST
(KINETIC ENERGY)
- **MECHANICAL**

INSIDE THE TUNNEL
IS A TURBINE
(PROPELLER)

LAKE

The faster the water (more energy), the faster the turbine spins.

153

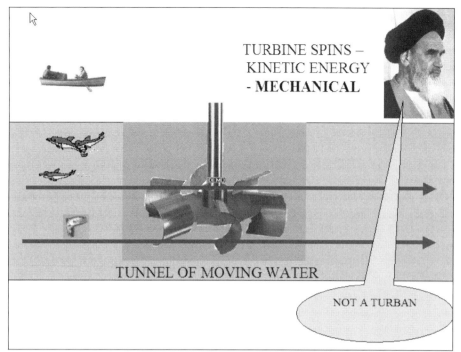

The turbine is connected to a generator, which also spins.

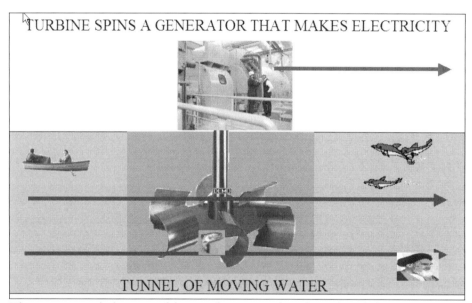

The generator is huge, as big as a house, and makes electricity. The electricity is sent to your house through power lines.

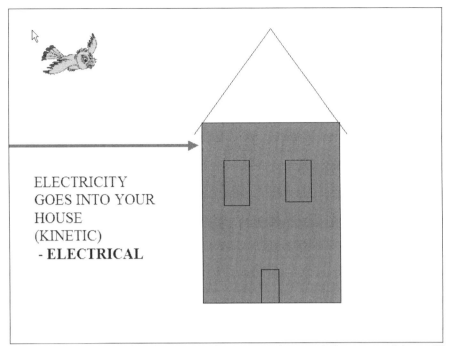

ELECTRICITY
GOES INTO YOUR
HOUSE
(KINETIC)
- **ELECTRICAL**

The electricity coming into your house can be used for many things, like watching the Philadelphia Eagles win the Super Bowl or an ugly face on TV.

AND INTO YOUR TV

Mike Ritts
wanted for
extream ugliness

reward

contact
junglecat3366@gmail.
com

caution
visually dangerous

The T.V. turns the electricity into light, sound and heat, which is absorbed by the walls of your house and changed into heat energy (infrared light).

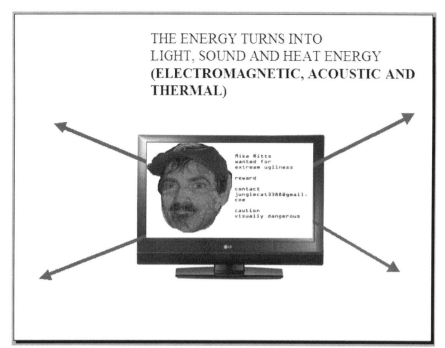

Your house then radiates this energy as electromagnetic energy (infrared)

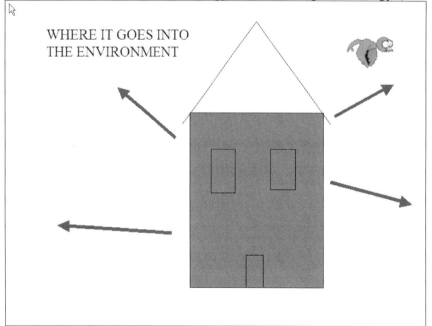

This energy radiates outward and eventually reaches outer space, going outward at the speed of light (186,000 miles per second).

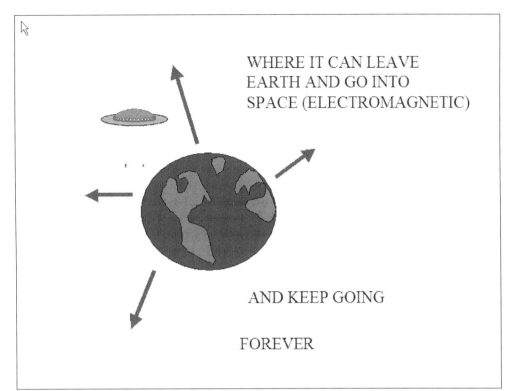

WHERE IT CAN LEAVE
EARTH AND GO INTO
SPACE (ELECTROMAGNETIC)

AND KEEP GOING

FOREVER

If aliens have radio telescopes (like we do) they can detect this energy and tell that earth has advanced life on it.

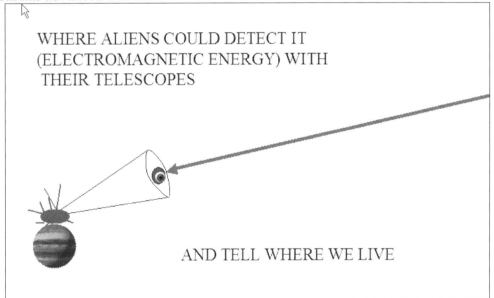

WHERE ALIENS COULD DETECT IT
(ELECTROMAGNETIC ENERGY) WITH
THEIR TELESCOPES

AND TELL WHERE WE LIVE

This is why humans are looking for intelligent life with radio telescopes. It is called the SETI project (search for extraterrestrial intelligence). We hope to detect an alien species this way. Of course, they could detect us instead and come and visit.

But happily:

In this example energy was converted from sunlight (kinetic electromagnetic energy) to rain (kinetic mechanical energy) to rivers (also mechanical energy) to a lake (potential gravitational energy), back to moving water (kinetic mechanical energy) to the spinning turbine (kinetic mechanical energy), to spinning the generator that made electricity (kinetic electrical energy) to your T.V., which produced light, sound, and heat. This was absorbed and radiated from you house as kinetic electromagnetic energy that went out into outer space. That was a long explanation and I am out of breath, but that is what happens in nearly every house in the USA.

Of course, we can spin the turbine other ways if we wish, wind can turn a turbine/generator, burning coal or gas can produce steam to turn the turbine/generator, even nuclear energy can make heat to boil water into steam and turn a turbine/generator. Even bad students could be used to turn a generator if needed! We would have to hook the generator up to a treadmill but it would work! Ok, maybe prisoners.

An interesting note:
On earth we have been sending electromagnetic radiation into space, in the form of Television since 1939. If there are any aliens within 75 light years or so, they could be watching our old T.V. shows! How is that for telling them about us! That is what SETI is looking for, only from other planets. Maybe they have T.V. too.

What about solar power

Solar power works a little different. A solar cell turns sunlight directly into electricity. There is a lot of sunlight but it is not very concentrated and not much sunlight is actually converted into electricity. But it is still a useful idea. I respect those who put solar panels on the roof of their homes but I am not sure it is economical for everyone.

Lost energy

Sometimes you hear that energy is lost. You should now realize that it is not lost; it is just not useful anymore. This brings us to the term *efficiency*; our machines are not as efficient as we might hope. Most of the energy we pay for is not used for its intended purpose; it is wasted, not lost. It turns into something not useful for our purposes, it is wasted, and we call this energy lost (to our use). Most of it turns into useless light, sound or heat. Think of a clothes dryer, it uses a lot of electricity to dry our clothes, but the darn thing is noisy, it may be unbalanced and jump around the room a bit, there is heat that goes out the exhaust pipe (that is the hose that goes out of your house) that is not used to dry your clothes. This is electricity you pay for but do not use for its intended purpose. Any energy you pay for that is not used to dry your clothes is considered "lost", or better yet not used for your benefit.

A TV or any other appliance is the same. If you look behind a television you may see light from the circuits or feel that it is warm, and maybe you will hear a slight humming sound, all wasted energy (which you pay for anyway). A car is the worst, the engine feels warm, there is light inside the engine where the gas is burned, and the car is noisy, all wasted energy, and this lowers your gas mileage, and costs you money.

Think of all the wasted energy we pay for. An air conditioner removes heat from your house and puts it outside. A refrigerator removes heat from inside (to keep your food cold) out the back. A light bulb is supposed to make light but feels warm when you touch it. A kitchen stove lets heat escape around the pot you are trying to heat. All of this is wasted energy. Cars are a good example, you pay for gasoline to make your car move, but only about 40% of that gas is actually used to move the car; the rest makes the car noisy, or hot, lost energy.

To be honest makers of modern appliances make an effort to make them more efficient. Cars can be made to get better gas mileage, televisions are made to be more efficient, so are water heaters, and refrigerators. There is a yellow tag on all appliances that tells you how efficient an appliance is. They are useful. Save energy and save money, everyone wins.

A day in the life of Earwig Hickson III

I was very worried about my Dad's health. Something was making him look old. I took it upon myself to make him healthy. The first thing I did was sneak onto the computer and buy a whole bunch of little generators with his credit card. It was so easy I decided to add a used radio telescope from NASA surplus, it was only $3.99 He would be upset when he got the bill but by then I would be a hero.

I started with an old exercise bike he never used. I had a plan. The big football game was on TV this weekend, and I was going to save him from himself. I hooked one of the generators to the exercise bike and ran the wires to the big flat screen TV he spent so much time in front of. I cut the original electric chord so the only way he could watch T.V. was by pedaling the bike and getting his needed exercise. I also noticed my Mom listened to the radio too much so I put a generator on an old elliptical for her. Since my sister spent so much time in the bathroom, I put pedals on the toilet to keep the light on; she always said she was busy, now she really was. I may have gotten a bit carried away with some things but I was trying to not only save my family but help save the world. I put a generator on every source of wasted energy I could find in the house. A generator went on the tetherball game, my pinwheel toy, and the exercise wheel for my hamster.

Then I turned my attention to the town. I had many more generators and many places to put them. I hit the playground first, turning kid energy into electricity. Generators went on the seesaw, the swings, and the merry-go-round. I went to City Hall where I had heard there was a lot of wasted energy. I could find nothing moving so I went on. The local creek got a water wheel hooked to one of my generators, and the escalator at the Mall. I put a treadmill in the park for people walking their dogs, and for joggers, each had a generator connected. I was making clean energy from useless energy, I was saving the world.

When I got home I hooked up the old Radio Telescope and sent out a message to space, hoping someone was out there, and was listening. I told them how wonderful Earth was and where I lived.

The consequences were not what I expected. My Dad missed the last 2 minutes of the big game when he nearly had a heart attack trying to keep up with the electrical demands of the Television. He missed the most amazing comeback in NFL history. My Mom nearly electrocuted herself trying to listen to the radio while taking a shower. My sister it turns out spends all her time in front of the mirror, not on the toilet; her make up was so bad she lost her date for the Prom. Evidently this is a big deal. It was not good for me either. The bathroom was dark because I could not pedal and "go" at the same time, and I am afraid of the dark. I also could not watch Television since my Dad would not pedal for me.

My punishment was not the usual grounding. This time I had to pay off my debt in Kilowatts of electricity. I was grounded until I produced 100 KW on my Dad's exercise bike. It would not have been so bad if he hadn't hooked that butt whacker up to it, as an added incentive. When the debt was paid I was in very good shape, but rather sore.

A side note: Thousands of years later Earth was in fact invaded by Aliens. They came straight to my address; the new owners of my house were surprised. It made the news. At least I got away with that one.

Review of terms – Quizlet:
https://Quizlet.com/124261755/chapter-14-volume-one-energy-transitions-flash-cards/?new

Fun things to Google

SETI project
Carl Segan's book – *Contact*.
How a generator works
How hydroelectric power is made
How electricity is made from coal
How electricity is made at a nuclear power plant
How a coal-fired power plant works –YouTube
How does a nuclear power plant work?

Links

Phet – how a generator works Faradays electromagnetic lab
https://phet.colorado.edu/en/simulation/faraday

Phet energy Skate Park
For real computers https://phet.colorado.edu/en/simulation/energy-skate-park
For chromebooks and tablets https://phet.colorado.edu/en/simulation/energy-skate-park-basics

Phet circuit lab https://phet.colorado.edu/en/simulation/circuit-construction-kit-ac-virtual-lab

Phet – energy forms and changes https://phet.colorado.edu/en/simulation/energy-forms-and-changes

Phet – Faraday's law https://phet.colorado.edu/en/simulation/faradays-law

Why haven't we found any aliens on other planets? Are we looking for it? What is SETI? An interesting film from Ted-ed. http://ed.ted.com/lessons/why-can-t-we-see-evidence-of-alien-life

Converting potential elastic energy into kinetic mechanical energy:
https://www.youtube.com/watch?v=2Vm1ZD1RjA8

Using hot steam to light a match and burn paper:
https://www.youtube.com/watch?v=73k9DuAS9v4

The gummy bear experiment – sugar is turned into heat and a cool fire:
https://www.youtube.com/watch?v=Ca4kY5gGI4c

The flour mill explosion – converting chemical energy into heat and a nice explosion:
https://www.youtube.com/watch?v=kRu0B1yKEl8

Converting wind energy into electricity with a generator:
https://www.youtube.com/watch?v=oBiGcZ5FmfU

Converting light energy into motion with a radiometer:
https://www.youtube.com/watch?v=CV_aGapVyaM

Hero's engine – converting heat into steam and into motion:
https://www.youtube.com/watch?v=_nvLzWH7-hc

Chapter 15
Energy Technologies

Movie of the power point for this chapter:
https://www.youtube.com/watch?v=9rq5qFRFstI

The wonders of science

The closest planet, the closest source of water, Oxygen and power was 250,000 miles away. That planet was Earth, the moon was much closer, but there was no help there. It was April of 1970; Apollo 13 was on its way to land on the moon. It was to be the third manned moon landing. Astronauts James Lovell and Fred Haise were to be the fifth and sixth humans to walk on the moon. Jack Swigert would stay in orbit around the moon.

Ten months earlier Neil Armstrong was the first human to walk on the moon. It was a big deal; it was televised live on TV, everyone watched. It was one of the greatest accomplishments of the human race, but that was ten months ago, landing on the moon was old news, no one cared about Apollo 13. Space flight was not interesting to the public anymore. NASA was boring.

The Apollo 13 spacecraft was powered by fuel cells. These are devices that combine Oxygen and Hydrogen gas into electricity and water. One of the Oxygen tanks exploded resulting in a shortage of Oxygen, water, and most importantly electricity. Without this electrical energy, the spaceship could do nothing. No computers, no heat, no coming back to Earth. Energy was the limiting factor, without it the three astronauts were doomed. In the end NASA had figured out how to make the little bit of electricity the craft had last until it returned to Earth and the Astronauts survived. It was an exercise in human ingenuity and creativity. It was about energy, and energy is very important.

Apollo 13, https://en.wikipedia.org/wiki/Apollo_13
The movie, http://www.imdb.com/title/tt0112384/

Energy Technologies

There is a conflict between the energy we need and the environment we value. It is not an easy choice, clean air, cheap energy. It is one that politicians struggle with all the time. Cheap energy means jobs and a good economy. Clean air means it is available for us (or more specifically you) in the next 50 years. Do you want a clean environment or cheap electricity? This is a serious question, and one not easily answered. I often think that young people, who cannot vote, should have a bigger say in this topic. The politicians who make these decisions do not expect to live more than 20 years or so, but you, the student, can expect to live for at least the next 60 years. You should have a say, but you do not. It is the reality.

The real reality is that in the United States of America we use (and need) a lot of energy. This is because our standard of living is very high, in other words we want this cheap energy. We need it, to makes our lives better. In America we have a lot of cars and they are big cars (SUV's, trucks, and minivans). We like air conditioning, our electronic toys, and heat in the winter; this is called our standard of living. It is not a bad thing, it is what we expect, and think we deserve. I am not here to take away your luxuries, I value them too. Rather I am here to help you make intelligent decisions about this topic.

What we need is a compromise, how to get what we want (luxuries) and what we need (a clean environment). It is not easy but the answer is there, *compromise*. This is where your opinion matters, are you more in favor of a strong economy (and jobs) or does the quality of the environment concern you more? It is a difficult question but one you (and all people) MUST address. As I said before, *compromise* is the answer.

So it comes down to what we can do to keep our toys and protect the environment also. The answer is not as hard as people pretend; it comes down to three things:

1. Efficiency

Conserve energy by using more efficient appliances. This means replacing old appliances with more modern versions like refrigerators, hot water heaters, television, stoves, and cars. The advancements are there, they usually cost more, but efficiency is what we are after. More bang for the buck, meaning getting what we want at a lower cost and lower environmental damage. Everyone wins. Our scientists and engineers are always improving the efficiency of appliances, cars get better gas mileage, and hot water heaters use less energy to do the *same job*. Modern flat screen television sets use less electricity to do the same job as the older "tube" sets. Light bulbs are more efficient today than 10 years ago; they still make the same amount of light but at a fraction of the cost. In America we value new technologies, pay attention, it works.

2. New sources

Finding energy from new sources is one possible solution. Coal is king in this country (slowly being replaced by natural gas), it is cheap and reliable, and most of our electricity is made from burning coal. I am not here to undervalue coal burning power plants, they do a good job. I am here to give alternatives. How about nuclear energy, not popular but clean (as far as air pollution). How about hydroelectric power (still has its disadvantages). How about solar power (still not perfect). Maybe wind power (not so fast, it has disadvantages too). The reality is there is *no perfect answer*; this is where your values and opinions come in.

3. Insulation

Adding insulation or energy efficient windows to your house is another option. It is expensive to install but you save on your energy bill for as long as you live in your house. The number one (by far) best way to save on your energy bill and help keep the environment healthy is insulation. As an added bonus it keeps your house more comfortable too.

It all comes down to the environment *versus* cheap energy. I make no judgment at this point. I just want you to have the facts, so you can make an intelligent decision on your own. There truly is no correct answer. It comes down to what you value more, the environment or your standard of living.

How do we get our electricity now?

Currently we get most of our electrical energy by burning fossil fuels (such as coal, oil and natural gas). These fuels are burned to produce heat, which boils water and makes steam; the steam is used to turn a generator. Unfortunately there are environmental costs to all of these. They produce carbon dioxide gas, which, although not poisonous, does contribute to global climate change. Global climate change may be a big problem in the future. Burning coal also produces chemicals that cause acid rain; this changes the composition of our farmland soil and slowly dissolves buildings and statues.

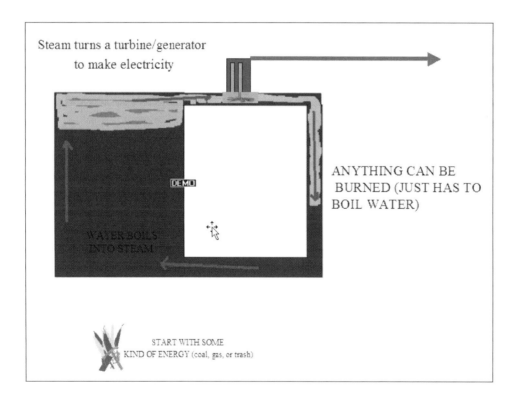

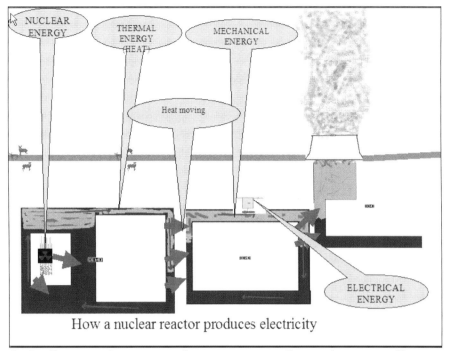

How a nuclear reactor produces electricity

So it all comes down to turning a generator. Currently we usually use a heat source to make steam to turn the generator, but other things can turn it as well. Wind power uses wind to make the generator spin, hydroelectric power uses river water and, nuclear power is hot so it makes steam to turn the generator. Tidal power uses the movement of ocean tides to spin the generator. All these methods have advantages and disadvantages

Other ideas we are trying are things like biofuels (made from corn), geothermal (using the Earth's heat) and solar (using the sun).

Advantages and disadvantages of energy sources:

COAL POWER OR NATURAL GAS

ADVANTAGE	DISADVANTAGE
CHEAP	AIR POLLUTION
DEPENDABLE	ACID RAIN
MAKE A LOT OF POWER	HEATS RIVER WATER
	GLOBAL WARMING
	NONRENEWABLE

NUCLEAR

ADVANTAGE	DISADVANTAGE
CHEAP	DANGER RISK
DEPENDABLE	
MAKE A LOT OF POWER	HEATS RIVER WATER
	NUCLEAR WASTE
NO AIR POLLUTION	
	NONRENEWABLE

BURNING TRASH

ADVANTAGE	DISADVANTAGE
CHEAP	AIR POLLUTION
DEPENDABLE	
GETS RID OF TRASH	NOISY
RENEWABLE	

WIND POWER

ADVANTAGE	DISADVANTAGE
CLEAN	EXPENSIVE TO BUILD
HELPS FISH HABITAT (IF IN OCEAN)	UGLY (TO SOME PEOPLE) NOT A LOT OF ENERGY (PER WINDMILL)
	NOT DEPENDABLE
RENEWABLE	KILLS BATS (PRESSURE CHANGES) AND BIRDS

HYDROELECTRIC

ADVANTAGE	DISADVANTAGE
CHEAP	CHANGE RIVER HABITAT
DEPENDABLE	
MAKE A LOT OF POWER	BLOCKS FISH MIGRATION
CLEAN	WARMS WATER ABOVE THE DAM
RENEWABLE	(NOT THERMAL POLLUTION)

SOLAR CELLS

ADVANTAGE	DISADVANTAGE
CLEAN	EXPENSIVE
USE IN RURAL PLACES	VERY WEAK POWER
RENEWABLE	NOT DEPENDABLE
	SHADE
	LAND USE-
UP TO10 ACRES PER
MEGAWATT

FUEL CELLS

ADVANTAGE	DISADVANTAGE
CLEAN	EXPENSIVE
NEW TECHNOLOGY
DEPENDABLE	LOW POWER
VERY EFFICIENT	HAVE TO BUILD ALL NEW
GAS STATIONS (HYDROGEN)
CAN TRIPLE CAR'S
GAS MILEAGE	EXPLOSIVE
	TAKES ENERGY TO
GET HYDROGEN

ELECTRIC (HYBRID) CARS

ADVANTAGE	DISADVANTAGE
CLEAN (AT CAR)	LOW POWER
EFFICIENT	EXPENSIVE
QUIET	USE ELECTRICITY FROM POWER PLANTS (SOME POLLUTION)
	NEW TECHNOLOGY

NEW LIGHT BULBS
LED BULBS

ADVANTAGE	DISADVANTAGE
VERY EFFICIENT	NEW TECHNOLOGY
CHEEP IN LONG RUN	EXPENSIVE TO BUY

INSULATION

ADVANTAGE	DISADVANTAGE
CHEAP IN LONG RUN	EXPENSIVE TO BUILD
DEPENDABLE	
SAVES A LOT OF ENERGY	

The environment

The environmental problems associated with making energy are complex too, mostly because politics have become involved. Looking at the T-charts above, you can see there is no perfect answer. I do not advocate for any one type of energy (although I do have my opinions); I just want you to make an intelligent decision about them. You are the ones who will be alive in 60 years, not me. Vote wisely.

Acid Rain

Acid rain is produced from burning gasoline in cars and coal in power plants. These release sulfur and nitrates that mix with rain to form weak acids in the clouds, which rain everywhere. The rain itself is not dangerous but over a long time is dissolves things out of the soil and changes the environment.

Global climate change

Something is definitely happening to our climate, it is probably from humans putting extra carbon dioxide (and other gases) into the atmosphere. For a long time this was no big deal but now we have reached a tipping point, the atmosphere (and the ocean) is kind of *full* after 150 years of this (since the industrial evolution). The extra carbon dioxide comes from burning fossil fuels such as gasoline, coal, and oil. The Earth is getting warmer, not much, only a degree or two, but this IS a big deal. You can expect in your near future, stronger hurricanes, bigger snow storms, more common droughts, and floods. This may sound surprising to some people, the earth warming up and getting more snow? This is about climate change not temperature. It is true that heat drives our climate and weather but it is not a warming, it is a change, an energy change. Climate change means deserts may form in our farm belts, storms will have more thermal energy and have more water vapor in them, including snow storms (from more evaporation), ocean levels will rise, at least a meter (probably more) which does not sound like much, but many of our cites are only a meter above sea level now. Most people in the world live on the coast.

Here is the situation; Global climate change could be good! But it may be very bad! It is a 50-50 shot, we do not know, we just know something will change. Since our climate is rather good right now, perhaps we should not take a chance. Imagine if the worst case comes true – a desert in the central United States, where we grow most of our food. Where will we get our food? Maybe New York City, and other costal cities, will be flooded, where will those people move? A lot of unknowns on this issue it is best to listen to the scientists (who know what they are talking about) than politicians who always seem to have an agenda.

So why do scientists think the climate is changing? Here is a short video from Ted-ed that goes into the reasons. http://ed.ted.com/lessons/rachel-pike-the-science-behind-a-climate-headline.

A day in the life of Earwig Hickson III

Maybe I am a little off, as my Dad says. Maybe I have a little too much imagination, as my mother says; maybe I am all the way crazy as my sister says. Either way I had a dream, more like a vision. In my dream I was in the future and I met me! I was leader of the world! I took over the world in my dream. I had been in charge for quite some time.

The other me told me all about the future, except the winning lottery numbers, which would have been useful. Anyway I took myself on a tour of the world. It was not good.

It was 75 years into the future and I got to see the cost of all the cheap energy we used. The Earth had in fact become warmer, only about 2 deg. C, but it made more impact than I ever expected. It was not necessarily hotter, but different, I mean the climate had changed, sea levels had risen a bit, some glaciers on land had melted and added extra water to the ocean, older me wanted to show me stuff, younger me was listening. Older me bought me a chocolate ice cream cone, how did he know I liked chocolate, oh wait, I know. So off we went on my tour. First we went to see the Statue of Liberty. She was still there but her island was gone. The great coastal cities were still there but no cars, only boats in the streets. New York City looked like Venice of old, Staten Island was the same. The ocean had only risen a few feet, but went inland quite far, and most people used to live (I mean live) just a few feet above sea level. The climates had shifted north; the great plains of the United States no longer produced the amount of food it used to. It does not rain as much in our farming belt as it once did. California is dry. We now import our food from Canada. Canada had to cut down its vast forest to produce this farmland, to feed us, and that made things worse. Older me tried to plant more trees, in the Midwest at one time, but it was too late, it was too dry. We went to the Pacific Islands of Indonesia, but there were none; they used to be just a few feet above sea level, not anymore. We visited the Mall in Washington DC. The Washington Monument was sticking out of a lake; it was cool, but in a bad way. The good news is that shipping lanes in the north were open that had always been iced-in before. There was now in fact a North West Passage! Henry Hudson was right, only a few hundred years too soon. The Earth was still here, the climate was still livable, but millions of people had to move to new locations, international borders were in the way. Many people's lives were completely changed. The economy was in chaos. It was a mess, an avoidable mess.

Older me, told me of mass food shortages, powerful hurricanes, bigger snowstorms, changes in rain patterns. He gave me a warning. You may have complaints about your weather at home, but they are minor, a few inconvenient storms compared to giant climate change are nothing. It is up to you, the younger generation, to make sure these changes do not come to pass. Plant more trees, cut down less, and use less fossil fuels and more renewable energy. Do not shrug off wind power, solar power, and tidal power, because it is more expensive. Do not let politics overshadow scientific data. Learn the facts; vote wisely, compromise between the economy and the environment. You will be taking over the world someday, he said, do not make the same mistakes you (I mean I) made. Think of the future, of your children, of your grandchildren, "I have children?" I said. He did not answer.

Then I woke up. Wow, what a dream, the kind that makes you think. I will be taking over the world someday, but it was only a dream after all.

Apparently, I had taken my camera along in my dream and when I woke up there were photographs on it, it was very disturbing, but of course it was just a dream…..

The photographs showed all the things my dream had shown me, I had a lot to think about, since I would be taking over the world soon, that is a great responsibility! I got right to work organizing my plan. I also planned to buy inland property as an investment, because I was sure it was soon to become valuable beachfront property. I would have rather had the lottery numbers though.

http://www.takepart.com/article/2015/07/15/what-sea-level-rise-looks-america-coast?cmpid=tp-fb

Review of terms – Quizlet:
https://Quizlet.com/124262746/chapter-15-volume-one-energy-technologies-flash-cards/?new

Fun things go Google
Global climate change
How electrify is made
Solar power
Solar cells
Fuel cells

Electric cars
LED light bulbs
Acid rain
Wind power
Bill Nye climate change
Neil deGrass Tyson climate change
Nuclear energy
Tidal power
Geothermal energy
How a generator works

Links

How do batteries work? A film by Ted-ed. http://ed.ted.com/lessons/why-batteries-die-adam-jacobson

Global climate change. http://ed.ted.com/lessons/rachel-pike-the-science-behind-a-climate-headline

Using a fuel cell to run an electric fan:
https://www.youtube.com/watch?v=ldvVpmzLdZs

Where does the Earth's energy come from? A short film from Ted-ed.
http://ed.ted.com/lessons/a-guide-to-the-energy-of-the-earth-joshua-m-sneideman

Why don't we use solar energy for all our needs? A film from Ted-ed.
http://ed.ted.com/lessons/why-aren-t-we-only-using-solar-power-alexandros-george-charalambides

Unit 5
Heat

Chapter 16
Temperature

Movie of the power point for this chapter:
https://www.youtube.com/watch?v=ISo2E1Xd1og

The wonders of science

I was hiking in Glacier National Park. It was summer and the weather was nice. I planned to walk around the Lower Two Medicine Lake, find a tributary, fish it down to the lake, and walk around the lake to the campground. A normal day, a normal situation, but it turned out bad. I found the tributary, I did my fishing, and I found the lake, but it was dark, I was not where I thought I was, I fell in the lake. No big deal at the time, I kept walking, but I was walking the wrong way. I realized this at 12:00 at night, but by then I was in trouble. I was cold. My body temperature was beginning to drop. My wet clothes were conducting my body heat away; I was in trouble and did not know it. When I realized I was lost, cold, and far from camp, I tried to build a fire, I failed. I wrapped up in my Mylar survival blanket (which saved my life) to wait for the morning. Fortunately for me, I was rescued at about 3:00 in the morning by my father, who for some reason, decided to circle the lake in his canoe to find me. I used the blanket as a giant mirror and used my last match, he saw me. I was saved. I was suffering from Hypothermia. My body temperature had dropped to nearly 90 deg. F. I was not thinking clearly. I was near death and had no idea of the problem I was in. My last rational thought was to light that match, when I heard him hollering for me. It was a close call.

The normal human body temperature is about 98.6 deg. F. If that temperature drops for some reason a person goes into Hypothermia. Many people have died when lost in the wilderness because of this, and it does not have to be cold for it to happen. Water is a good conductor of heat so a person who is wet will lose body heat faster than they can produce it. In cold weather it is worse, some people call it freezing to death, but that is not really accurate. It is whenever your body temperature drops below normal and you cannot think right. Some people have been known to remove all their clothing when in a state of hypothermia, they became very irrational.

Hypothermia, https://en.wikipedia.org/wiki/Hypothermia

To build a fire (Jack London), https://en.wikipedia.org/wiki/To_Build_a_Fire

Temperature
Always remember and never forget

Matter is made of **moving** particles. All particles (molecules) in matter have some energy, so they move or shake. Gases move the fastest (fly around), liquids slide around, and solid shake. They all move though, as long as they have energy, and they all do have energy, at least above absolute zero temperature.

The kinetic theory of matter

The **kinetic theory of matter** simply says that all particles (molecules and atoms) that make up matter are in *constant motion*. Solids vibrate, liquids slide around and gases fly around. These particles are always moving. The faster they move, the *more* energy they have. This energy is *kinetic thermal energy*.

Phet kinetic theory of matter - https://phet.colorado.edu/en/simulation/gas-properties

Thermal energy

Thermal energy is the *sum* of how fast **ALL** the molecules in an object are moving. It is actually how much *kinetic energy all the particles in an object have*. Big things (with more molecules) have more thermal energy than little objects at the same temperature, because they have more molecules (the total is bigger). Even if the small thing is hotter, a big thing, which is "cold", may have more total thermal energy. For example the whole ocean at 60 deg. F has way more energy than a cup of boiling water (212 deg F), because it has zillions of more molecules.

This is why small ponds freeze in the winter before larger lakes. The larger mass of the lake has much more thermal energy because it has many more vibrating molecules; it takes a long time to lose that thermal energy. The small pond though has many fewer vibrating molecules and loses its energy much quicker.

A simple experiment you could do to show this is to fill your bathtub with hot water, and a cup with the same temperature water. Measure the temperature every few minutes. You will find that the cup of water cools down much faster than the bathtub. This is because the bathtub has more molecules thus more thermal energy than the smaller cup. (For an excellent explanation of this go to *Eureka episode 21 temperature vs. heat*. https://www.youtube.com/watch?v=AqDsAVPjgS4)

The important thing for now is that *thermal energy does not move*, it stays in the object.

In summary:

THERMAL ENERGY

- THE SUM OF HOW FAST THE MOLECULES ARE MOVING
- **BIG** THINGS HAVE **MORE** THERMAL ENERGY THAN SMALL THINGS (OF THE SAME TEMPERATURE) BECAUSE THEY HAVE MORE MOLECULES
- EVEN IF THE SMALL THING IS **HOTTER**
- HOW MUCH KINETIC ENERGY THERE IS IN THE **WHOLE OBJECT** (OR SYSTEM)

Imagine 2 cup of equally hot coffee over two campers' heads and they both spill. Which will burn for the *longest*?

The big cup has more molecules, thus more thermal energy. The little one cools off faster.

Heat

So what is heat? It is not thermal energy which is the energy in the *whole* object. **Heat** is when thermal energy *moves*. Thermal energy cannot really hurt you, even if it is a chunk of super hot metal, it is harmless to you, there is no way for that energy to get to you, unless you *touch* it, then it can harm you. When you touch the hot object the thermal energy moves *into* your hand and burns it. This is heat, the *movement* of thermal energy. Heat can hurt you. Heat always moves from hot (high temperature) objects to cold (low temperature) objects. Your hand was colder than the hot metal, so it moved into your hand. OUCH.

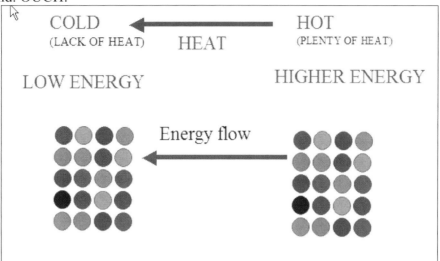

In the winter since your body temperature is higher than the outside air, the thermal energy leaves your body as heat and you feel "cold".

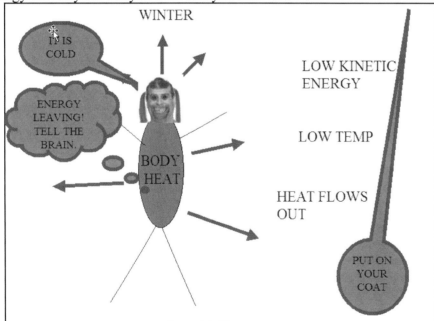

Insulation is a material that slows down the movement of heat, so you feel warm when you wear a coat.

FLUFFY
GREEN
EAGLES
JACKET

HEAT GETS
TRAPPED INSIDE

BODY
HEAT

In the summer your body temperature is lower than the outside air and heat moves into your body. You feel hot.

SUMMER

HEAT IS
COMING
IN

TELL BRAIN
TO TAKE OFF
CLOTHES

A LOT OF
KINETIC
ENERGY

HOT

I AM NOT
WEARING
ANY
CLOTHES

Temperature

What is temperature, it is not heat, and it is not thermal energy, what is left? **Temperature** is the *average* kinetic energy of an object at one location. Temperature is a *measurement* of how much thermal energy is present. A thermometer is used to measure how fast the molecules are moving. It assigns a number to the kinetic energy (thermal energy) of an object. Remember hot molecules move faster and cold molecules move slower. A thermometer measures this.

It is interesting for me how a thermometer measures this. It is by *thermal expansion*. When molecules move faster, they bump into things harder and make the whole object expand a bit (or get bigger). The liquid in a thermometer does this when you put it in a hot liquid; it expands and fills a little tube, with numbers on it. This is what a thermometer is, a little tube with numbers that allows liquid inside to expand and contract.

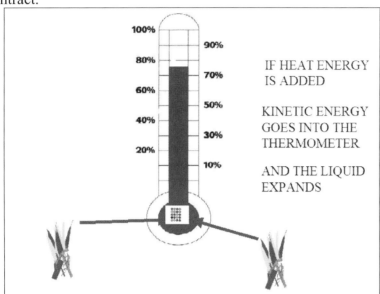

When heat goes into the liquid in a thermometer it expands and goes up the tube.

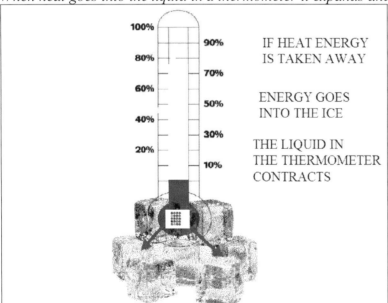

When heat leaves, the liquid in the thermometer shrinks and goes down the tube.

Types of thermometers

Although all thermometers work the same (thermal expansion) there are different types, the only difference is the numbers. Fahrenheit is used in the United States (we are used to it); Celsius (or centigrade) is used by everyone else. There are only three conversion numbers you really need to know, the freezing point of water, the boiling point of water and room temperature.

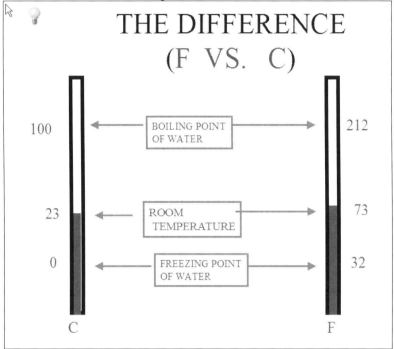

There is another kind of thermometer that you have never heard of but actually makes the most sense, a **Kelvin** thermometer. This one starts at the lowest possible temperature, absolute zero (where there is NO thermal energy), and goes up from there. It really shows that when I say it is "cold" there is still a lot of thermal energy present. It can never be below zero Kelvin, because nothing can be colder than "no heat').

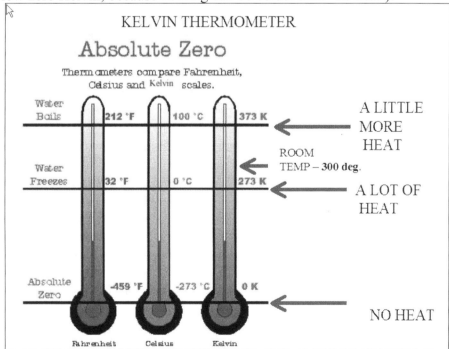

In a way, it is kind of cool to say, "My, this is a nice day, the temperature is only 300 deg. K, just perfect."

In review

COMPARE

THERMAL ENERGY	HEAT	TEMPERATURE
HOW FAST PARTICLES MOVE	MOVEMENT OF ENERGY	MEASUREMENT OF ENERGY
TOTAL AMOUNT OF KINETIC ENERGY	FROM HIGH TO LOW KINETIC ENERGY	HOW MUCH KINETIC ENERGY ON AVERAGE
BIG THINGS HAVE MORE KINETIC ENERGY	CAN FEEL	

Thermal expansion

This is a good time to introduce thermal expansion, since this is how a thermometer works. In a nutshell, hot things get bigger, and cold things get smaller. Hot things expand and cold things contract. This is because as molecules lose energy and stop hitting each other as hard, they get a little bit closer together; the whole material shrinks a tiny bit. As things get hotter the molecules move faster and hit each other harder, the whole object gets a tiny bit bigger. Metals expand (a little) when they are hot; gases expand (a lot) when they get hot, and so do liquids (just a little).

Well, not all liquids, water is special, it is unique. A glass of boiling water at 100 deg. C will in fact get a bit smaller as it cools, but when it hits about 4 deg C above freezing it does something odd, it begins to expand. It expands a lot too, about 10%. This is why ice floats on water (less dense) and why water pipes and garden hoses break when the water inside freezes (it expands). Make sure the pipes in your house never freeze (or they will break) and to empty the water out of your garden hose before winter (it will break if you do not).

But all other materials get smaller as they get colder. This is why bridges have expansion joints in them, so they can stretch and shrink with temperature. This is why sidewalks have lines cut into the cement, so the sections do not brake when they expand and contract. This is why you can remove a tight metal lid from a glass jar by holding it under hot water for a while; the metal expands more than the glass. You can even remove a tight ring from your finger with soap and warm water.

184

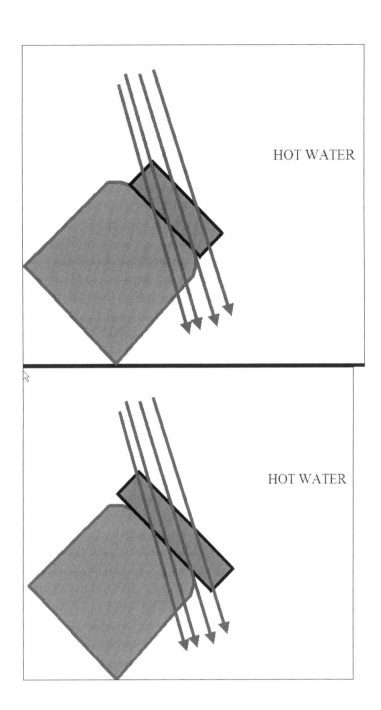

HOT WATER

HOT WATER

185

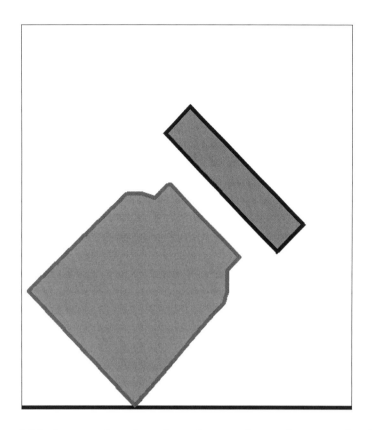

This is a good explanation of expansion and contraction by Eureka:
https://www.youtube.com/watch?v=rMg1bmF5BVo

Bridges are big, and a little expansion (or contraction) of the metal can make a big difference. The bridge could actually break and fall down. To prevent this, builders put expansion joints at the ends so the expanding materials have somewhere to go. This is how it works.

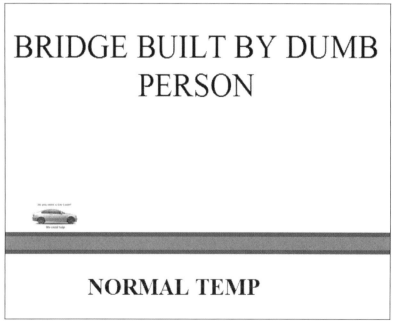

186

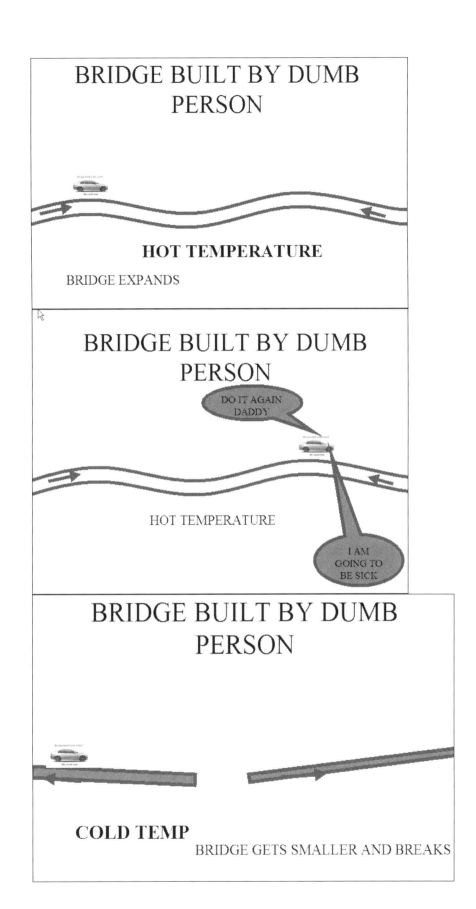

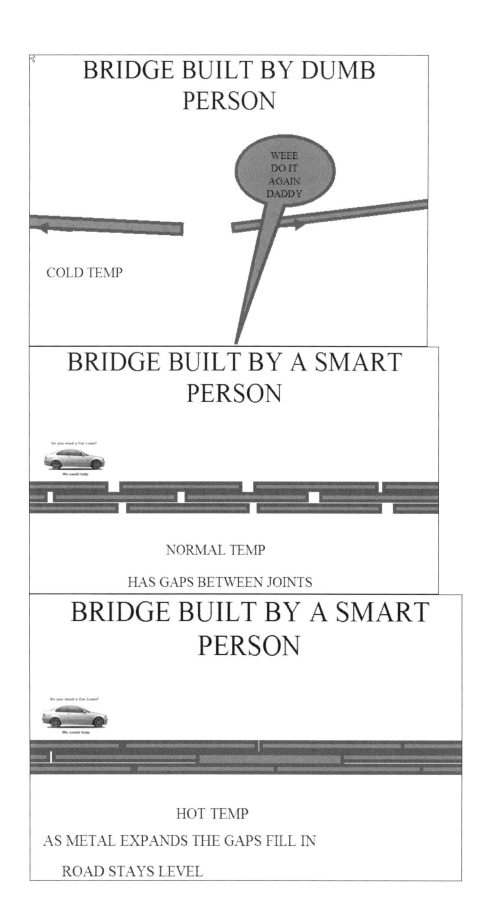

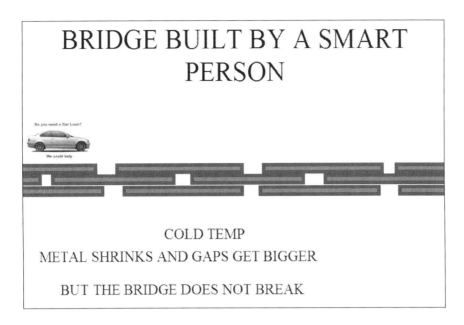

BRIDGE BUILT BY A SMART PERSON

COLD TEMP

METAL SHRINKS AND GAPS GET BIGGER

BUT THE BRIDGE DOES NOT BREAK

How can thermal expansion be useful?
Systems and feedback – a closed loop system

It turns out that even though all materials expand when they get hot, they expand at different rates. This can be used for a simple feedback device called a thermostat. Feedback is when part of a system gives useful information to the whole system so it knows what to do. In the case of a thermostat it tells the heating and air-conditioning when to turn on in your house. If it gets too cold, the heater turns on, if it gets too hot, the air conditioner turns on. It is just a matter of setting the temperature you want and the system works perfectly from there.

The secret to a thermostat is a piece of bimetal, which is two kinds of metal fused together (one on each side). These metals are usually iron and brass. When it gets hot, the brass expands more than the iron causing the whole piece of metal to bend in one direction, when it gets cold, brass shrinks faster and the metal bends the other way.

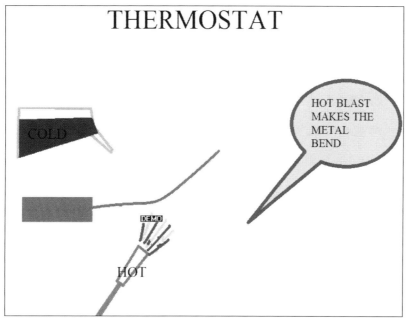

Heat makes the bimetal bend up.

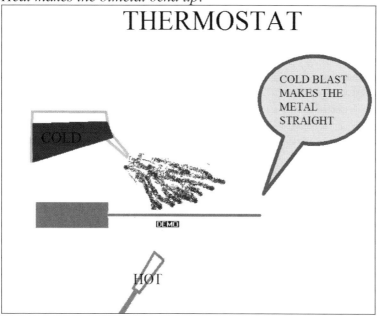

When it gets cold the bimetal bends down and becomes straight again.

The thermostat (bimetal) is the feedback mechanism. It is part of a larger system that controls the temperature of your house. This is part of a closed loop system, which is a system with feedback. This is how it works.

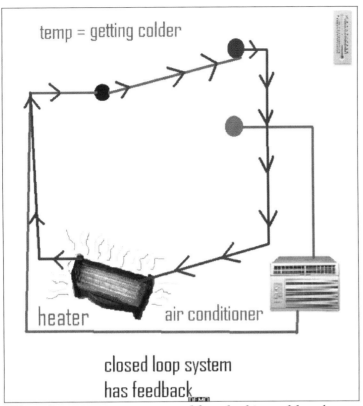

When the temperature gets colder, the bimetal bends upward and completes an electric circuit that turns on the heater

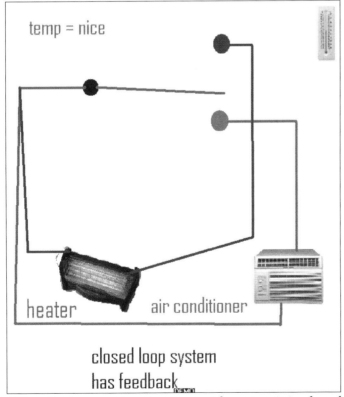

When the temperature is nice, neither circuit is closed, and both the heater and air conditioner are off.

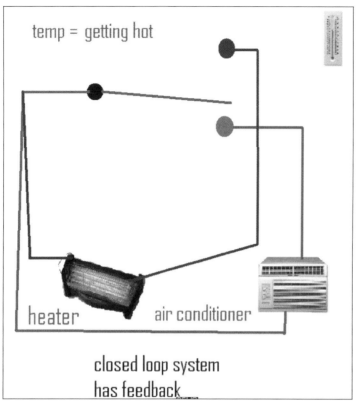

As it gets hotter the thermostat bends towards the air conditioner.

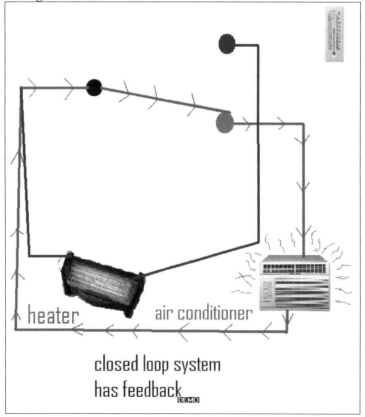

The air conditioner comes on.

192

Any system that gives feedback is a closed loop system. Any system that does not give feedback is an open loop system. An open loop system would be something like a lawn sprinkler that is controlled by a timer; the result is that the sprinkler will come on everyday at the same time whether it is raining or not.

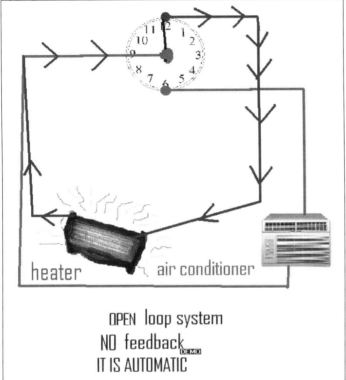

This heater comes on at 12:00 no matter what the temperature is. It could be 100 deg. F and it would still go on. It is an open loop system, and has no feedback to tell the heater what to do.

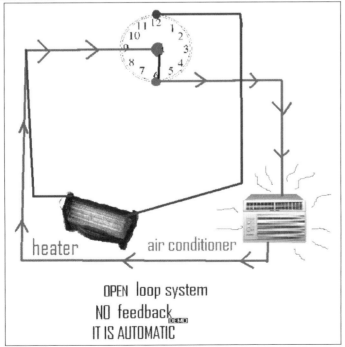

The same system causes the air conditioner to come on at 6:00, no matter what the temperature is. No feedback means it is an open loop system. The temperature could be below freezing and the air conditioner would still come on.

A day in the life of Earwig Hickson III

My Dad is an expert at feedback. It is not always positive but it is always helpful. When I do good things, he gives me feedback in the form of praise or better yet, MONEY! When I do something bad, he gives me feedback also, so I know it was wrong. The feedback he gives me is "useful", it gives me valuable information, and it helps me succeed. Since I want him to pay for my college education, his feedback is valuable to me. I am a *closed loop system* because I get feedback.

You see, I am a system, just like any machine, such as my bike. He knows that if one part of the system fails, the whole system fails. He does not want me (the system) to fail, so he is always sure to give me *valuable feedback*, it helps me, it helps the system.

Imagine my bike, a simple system of many parts. If the chain falls off, the whole system fails; I may as well have NO bike. All the parts of the bike are needed to make the system work. Some systems, like a bike have no feedback (information going back to the system), and when the chain falls off there is no information for the system to fix itself, it just stops. My bike is an *open loop system* since it gets no feedback.

When I mess up (and all of us do), the feedback I get from my Dad lets me know I messed up, I now know not to do that anymore. My system is still working efficiently; the feedback just lets me know what to change. For example I now know it is not good when I interrupt his football game, when there is only two minutes left, he lets me know this with useful feedback, and I know not to let it happen again!

When a system with no feedback fails, the whole system just stops, a total fail. But some systems, like me, get useful feedback and can repair breakdowns and keep the system working.

One of my favorite systems is my *sister control device*. I like when she stays in her room and does not bother me, so I built a system to keep her there. It starts with a garden hose pointed at her bedroom door. A motion sensor can tell when her door moves. When it opens, it sends a message to the water hose to turn on. When the door closes again, the motion sensor sends a message to the water hose to turn off. I do not have to do anything; the system takes care of everything.

Sadly, my Dad did not like a water hose in the hallway and gave me some useful feedback, so I do not do that anymore. I ended up in the corner, supposedly thinking about how I should love my sister, but in reality I was designing my new closed loop feedback system that would squirt slime all over my sister when she left her room.

Review of terms – Quizlet:
https://Quizlet.com/124264242/chapter-16-volume-one-temperature-flash-cards/?new

Fun things to Google

How a thermostat works
Bridge disasters
Eureka episode 21 temperature vs. heat
Insulation
How a thermometer works
Absolute zero
Liquid nitrogen
Thermal expansion
How to get a tight ring off your finger
How to get a lid off a jar
Eureka expansion contraction

Links

A trick using the thermal expansion of air to put a water balloon in a bottle:
https://www.youtube.com/watch?v=6JBIvlFvw-4

Bimetal bending and unbending with heat:
https://www.youtube.com/watch?v=RWsklayjdzo

An aluminum foil bridge expands with added heat:
https://www.youtube.com/watch?v=mtn26Tqe0-w

Ring and ball – using thermal expansion to make a ring bigger:
https://www.youtube.com/watch?v=Liz3ziMplic

More about feedback and closed loop systems. A film by Ted-ed.
http://ed.ted.com/lessons/feedback-loops-how-nature-gets-its-rhythms-anje-margriet-neutel

A magic trick using a hot piece of bimetal. It should be no secret to you at this point.
https://www.youtube.com/watch?v=C0HMbtIPHVA

Chapter 17
Heat

Movie of the power point for this chapter:
https://www.youtube.com/watch?v=uMA2UhTUT3Y

The wonders of science

It all started with cannons in 1798. Heat was once thought to be some kind of fluid that flowed from hot things to cooler things. There was only thought to be so much heat and it could not be created (or destroyed for that matter), it just flowed from hot objects to colder ones. This is where the cannons came into play.

A man named Benjamin Thompson (Count Rumford) noticed that when the factory he worked at, drilled the barrel of cannons, they got hot. What he noticed was that two "cold" objects, the metal cannon and the metal drill became hot with no apparent source of "heat fluid". He found that this heat could be made to boil water. This experiment proved that heat could be "created" from no heat and the heat produced had no limit. The heat was actually made from the friction and motion of the cannon and the drill. This was the beginning of the Kinetic Theory of Energy, in other words, motion could be transferred or changed into heat energy. You can test this yourself by rubbing your cold hands together and heating them up. It is a wonder no one noticed this before Count Rumford.

A short video by Eureka describing all of this:
https://www.youtube.com/watch?v=o3gN9wI_w64

Heat

At this point you should know that thermal energy is the total amount of energy in the whole object, but heat is when that energy moves. Thermal energy cannot hurt you but heat moving into you can.

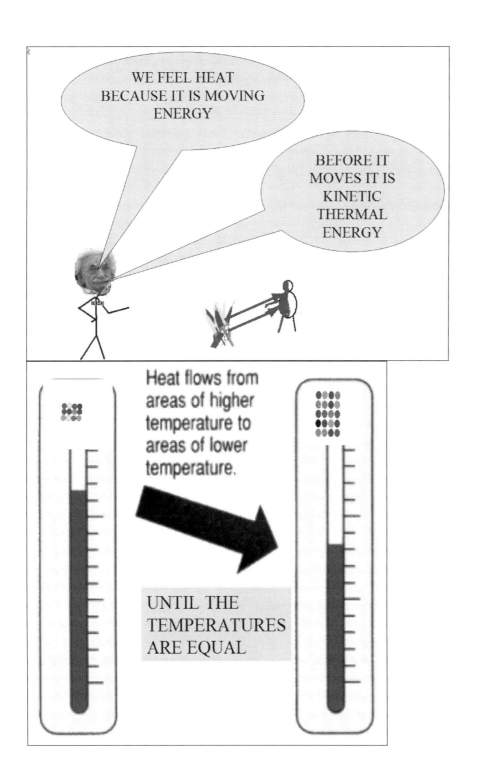

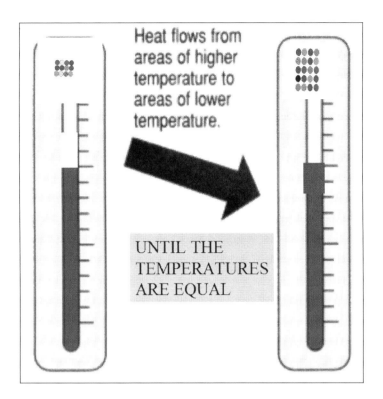

Heat flows from areas of higher temperature to areas of lower temperature.

UNTIL THE TEMPERATURES ARE EQUAL

This is going to be a sticky nasty chapter. This is the hard stuff, but since you are getting smarter now, it should be OK. At least I hope so.

Calorie

You have heard this word when talking about food and the calories it contains. Calories are potential heat energy in something, but unfortunately you learned this all wrong so let's start over.

A **calorie** is the *amount of heat energy needed to raise the temperature on one gram of water one degree Celsius*. **It is very important to remember this definition only applies to water**. It is a unit of heat energy. This is not very much energy. One gram of water is the same as a milliliter, and 5 ml would fit in a teaspoon. The amount of energy needed to raise the temperature of this small amount of water only one degree C. This is a science calorie (spelled with a lower case "c". A short film explaining a calorie: https://www.youtube.com/watch?v=XJRjV-aIaSY

A food calorie (what you are familiar with) is actually 1000 science calories or a Kilocalorie, or sometimes called a K-calorie or a large calorie. When people talk about food calories they use a capitol "C", as in Calorie. So one Calorie in a piece of candy is actually 1000 science calories. One food calorie would raise the temperature of 1000 grams of water one degree or one gram of water, 1000 degrees! This is 1000 times more energy than a science calorie. Burning food Calories is why you have a high body temperature; it makes a lot of heat. It also allows you to move around and do all the other life processes. An average person eats about 2400 food Calories per day, which is a lot of energy, enough to raise a 2 liter soda bottle full of water 1200 degrees Celsius. Unbelievable, that is a lot of energy.

"Want to lose 1200 Calories a month? Drink a liter of ice water a day. You burn the energy just raising the water to body temperature."
Neil deGrass Tyson

Do you know how scientists know how many calories are in a potato chip? They burn it, and use it to heat water in a special instrument called a calorimeter. After the food has been burned and the temperature change measured, the number of food Calories can be calculated. If you want to see how much energy is actually in food, just burn a Doritos chip or a peanut, it will burn for a long time. In fact a bag of Doritos can be used to start an emergency campfire on a rainy day. A good survival tip.

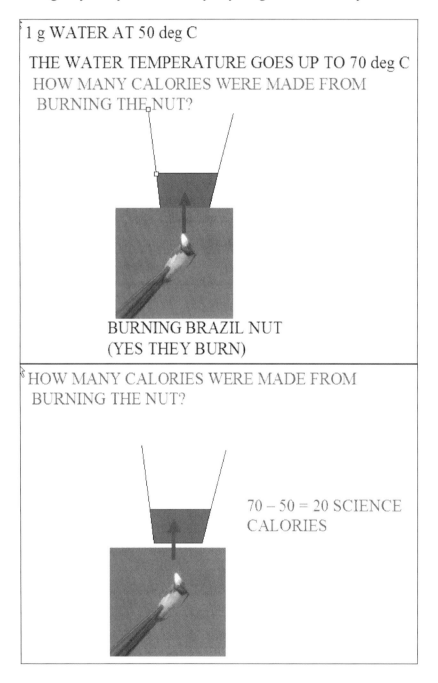

1 g WATER AT 50 deg C

THE WATER TEMPERATURE GOES UP TO 70 deg C
HOW MANY CALORIES WERE MADE FROM
BURNING THE NUT?

BURNING BRAZIL NUT
(YES THEY BURN)

HOW MANY CALORIES WERE MADE FROM
BURNING THE NUT?

70 – 50 = 20 SCIENCE
CALORIES

Specific Heat capacity

This is the hard part. A calorie is how much heat it takes to raise a gram of water one degree Celsius. **Specific heat capacity** *is how much heat (how many calories) it takes to raise one gram of a material one degree Celsius*. Every kind of material has a different specific heat capacity. Specific heat capacity really means how much heat a material can absorb before it starts getting hotter; it is how much thermal energy a material can hold. Some things heat up very fast when you put them in a fire (like metal); this means they have a low specific heat capacity. Other materials, like water, heat up very slowly in the same fire, and have a high specific heat capacity. Since water has such a high specific heat capacity (and can hold A LOT of heat) it is used to cool things down quickly. https://www.youtube.com/watch?v=k4N7zms21fo. Put a super hot piece of metal in water and it gets cold fast, the water absorbed all the extra heat. Throw water on a campfire and it goes out and gets cold. Throw metal on a fire and that does not happen; metal does not hold much heat before it gets super hot.

Substance	c in cal/gm K or Btu/lb F
Aluminum	0.215
Bismuth	0.0294 — HOLDS THE LEAST HEAT
Copper	0.0923
Brass	0.092
Gold	0.0301
Lead	0.0305
Silver	0.0558
Tungsten	0.0321
Zinc	0.0925
Mercury	0.033
Alcohol(ethyl)	0.58
Water	1.00 — HOLDS THE MOST HEAT
Ice (-10 C)	0.49
Granite	0.19
Glass	0.20

The heat capacity of some materials

So water happens to have a high specific heat capacity, and it takes a whole science calorie to raise the temperature of one gram of water one degree C. Bismuth (a metal) has a specific heat capacity of only 0.0294 calories. This means that the same calorie of heat needed to raise a gram of water one degree will raise one gram of Bismuth 34 degrees! The Bismuth cannot hold much heat, so it fills up quickly and begins to rise in temperature quickly. If you try and boil a pot full of water, it takes a long time, because water absorbs and holds so much heat.

CALORIE	SPECIFIC HEAT CAPACITY
ENERGY TO RAISE ONE GRAM OF **WATER** ONE DEGREE	ENERGY (NUMBER OF CALORIES) TO RAISE ONE GRAM OF **ANYTHING** ONE DEGREE
UNIT OF HEAT	HIGH VALUES HOLD A LOT AND HEAT UP SLOWLY
WATER HAS A SPECIFIC HEAT OF ONE	LOW VALUES DON'T HOLD MUCH AND HEAT FAST
	A PROPERTY OF A SUBSTANCE

The campers

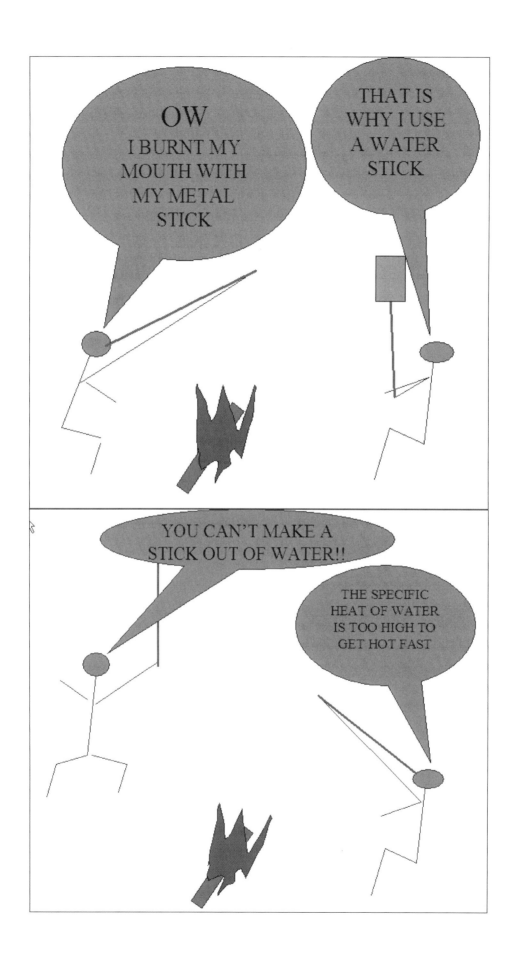

Things with low specific heat capacity – like metal:
Heat up very fast
Cool down very fast

Things with high specific heat capacity – like water:
Heat up very slowly
Cool down very slowly
A short video describing specific heat capacity:
https://www.youtube.com/watch?v=Kc7TKzUys3o

This is why people who live near the ocean have moderate weather compared to people who live inland. The huge amount of heat (thermal energy) in the ocean water keeps the land near the coast warmer. The rocks and dirt inland cannot hold as much and lose it quickly; the people living there have a colder climate.

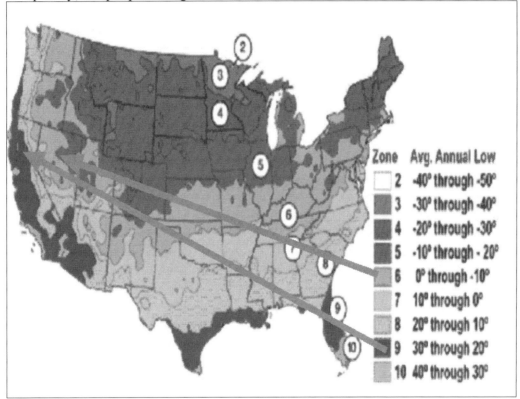

Notice the coast of California is warmer than Nevada because the Ocean helps keep it warm.

The bathers

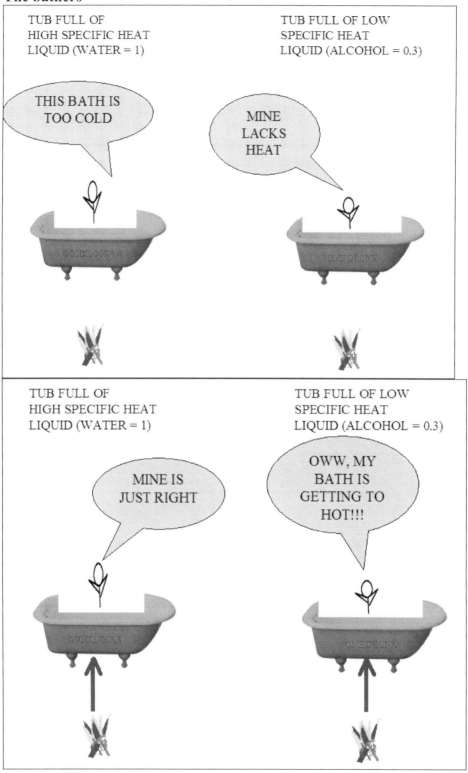

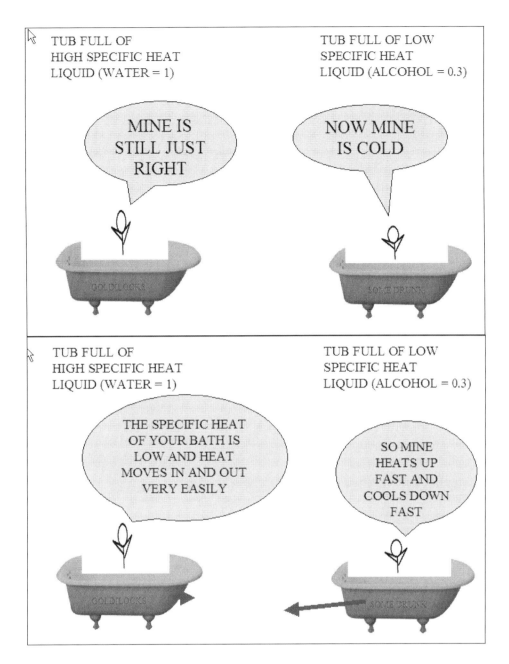

If you put a metal poker into a hot fire it will begin to glow bright red. This is because it can't hold much heat; in this case it transfers the extra heat into light, which makes it glow.

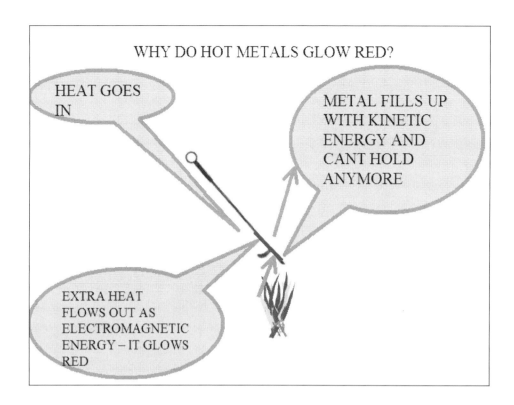

A story about snakes

Snakes have a tendency to hide from the hot daytime sun under shady rocks, but at night it gets too cold under there so they come out and lay on top of the warm rocks. Snakes are not dumb (or slimy for that matter). This is how it works.

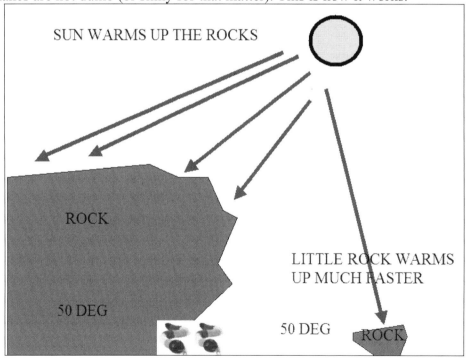

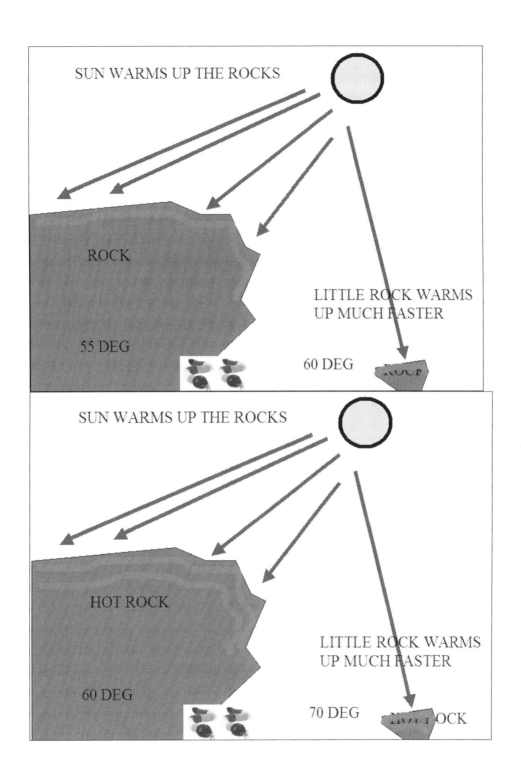

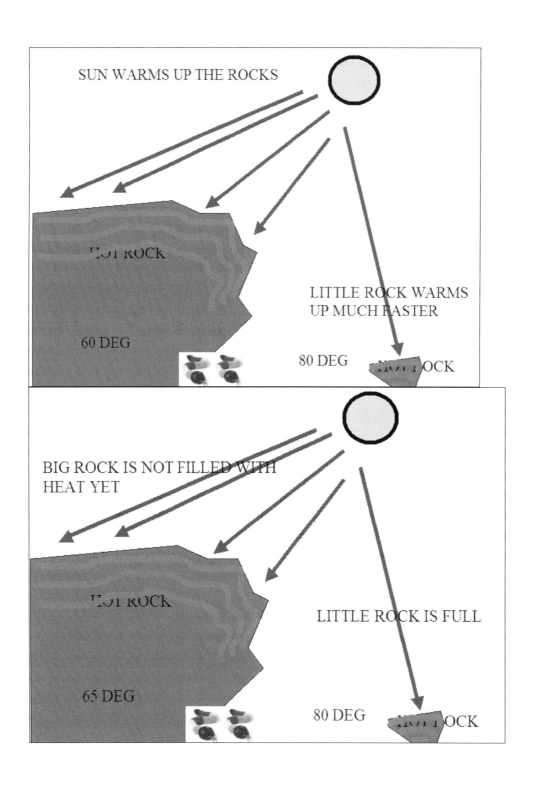

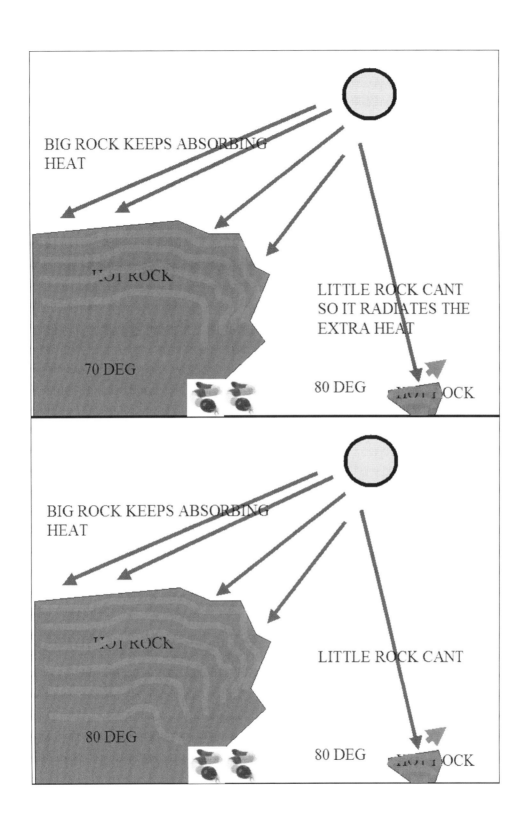

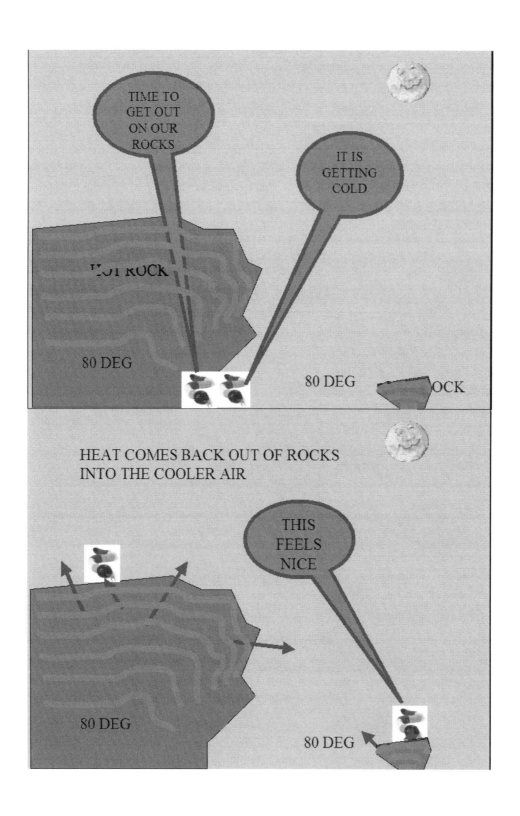

212

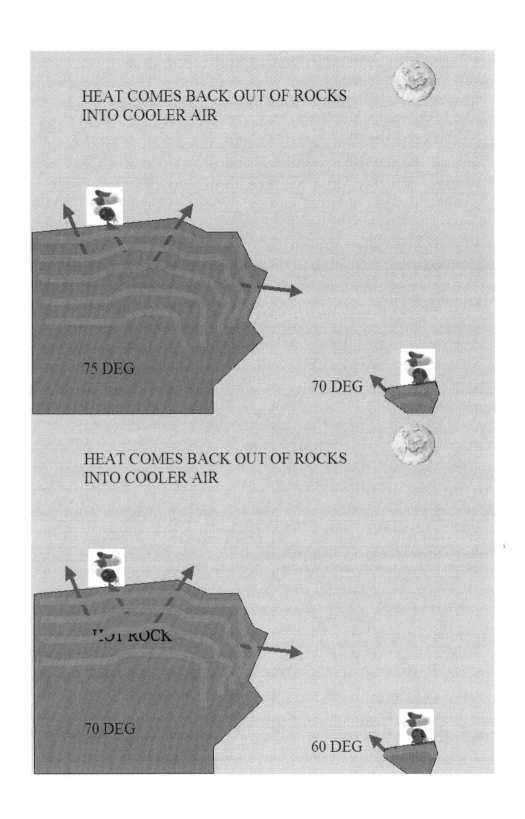

213

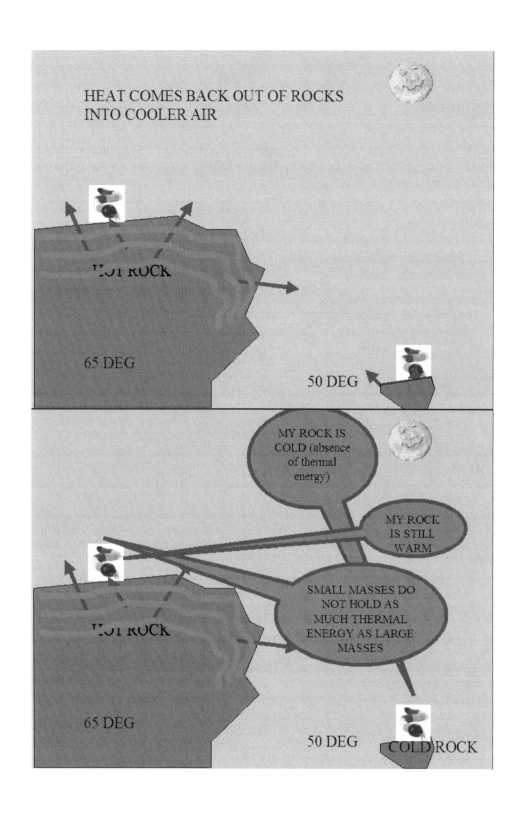

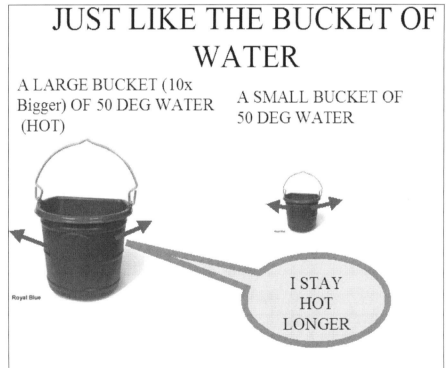

A day in the life of Earwig Hickson III

My bedroom is cold, my dad believes in saving energy (so do I) but he saves it by turning off the heater in my bedroom. This is not good, I get cold. I decided to do something about it. The first thing I did was throw away my old bed, it was not helping, out the window it went and I used it as a trampoline. The second thing I did was get a whole lot of rocks. I made my new bed out of the rocks. That afternoon I built a small fire under my new bed to heat up the rocks, I put it out just before bedtime, the rocks heated up very fast. It made my room a bit smoky but the rocks got very hot. That evening I crawled on my new bed to go to sleep. It was very hot at first, and I could not sleep, but soon it was very comfortable and I was soon snoozing. The rocks had released their heat to me and made me very comfortable. About midnight I woke up, the rocks were freezing, and I was freezing, the thermal energy was gone, and sleeping on rocks is not fun. I had to improve my bed.

I found an old waterbed mattress at a yard sale and put that on top of the rocks. Once filled with water I built the fire underneath and began to heat not just the rocks but the water too. It took a long time for the water to get hot. Finally my bed was ready and I snuggled in. It was toasty for the whole night. The water had held the heat and the heat moved into me and kept me warm. Success!

It turns out that water has a very high specific heat capacity and takes a long time to warm up, but once it is warm, it stays warm. The rocks with a low specific heat capacity heated up fast but only stayed warm for a short time. I got to thinking about this. I do not like cold, and in the winter it is cold. I had learned that large bodies of water like oceans keep the land warm nearby during winter. I decided to make my own lake. I dug a huge hole around our house and began filling it with water. It took a lot of water and a lot of time. Since there was so much water, it held a lot of thermal energy and it would keep my house warm all winter!

Sadly not all was perfect. My dad was not happy when he got home and found a moat around the house. He tried to drive through it thinking the driveway still had to be there, it wasn't, the car sank. On top of that I almost burned down the house when I built the fire under my bed. It was a very unhappy evening.

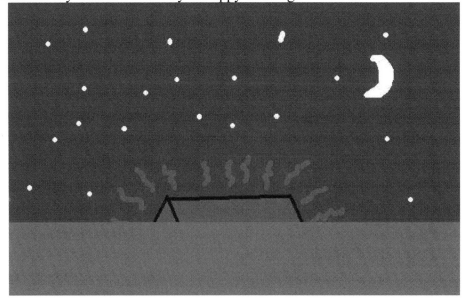

Review of terms – Quizlet:
https://Quizlet.com/124265167/chapter-17-volume-one-heat-flash-cards/?new

Fun things to Google

Why are snakes most active in the morning?
Calories in junk food
How do branding irons work?

Links

Specific heat with Napoleon and Newton
https://www.youtube.com/watch?v=Kc7TKzUys3o

What is a calorie? Where does the energy from food go? A short film by Ted-ed.
http://ed.ted.com/lessons/what-is-a-calorie-emma-bryce

How to boil water in a balloon: https://www.youtube.com/watch?v=qeDZQ9-gsjY

Conduction of heat down a copper bar:
https://www.youtube.com/watch?v=6hxLsqkgp7E

Balloons filled with air and water are placed above a candle:
https://www.youtube.com/watch?v=hot0ln4I-UE

A neat trick showing how a full can of soda can prevent paper from burning:
https://www.youtube.com/watch?v=bFDfdzJg6dg

A copper plate keeps paper from burning, a cool trick:
https://www.youtube.com/watch?v=P1gpFV5tzxA

Chapter 18
The Movement of Heat – Controlling Heat

Movie of the power point for this chapter:
https://www.youtube.com/watch?v=bYvuacP6n9A

The wonders of science

They did it in 1909, 68 men, and 246 dogs. They did it on sleds. They did it without snowmobiles, they did it without hand warmers or gas stoves or even tents. They were the first men to reach the North Pole. Admiral Robert Peary gets the credit (debatable) but it took an expedition to get him there. It was a trip over sea ice in the most inhospitable climate imaginable. The average temperatures were an unimaginable, - 40 deg F, they wore deerskin parkas, bearskin pants and sealskin boots to try and stay warm. They slept in igloos.

The secret to a successful trip depended on controlling heat. These people had to keep warm in an environment that tried its best to freeze you. Exposing skin to these elements resulted in body heat radiating away resulting in frostbite (freezing skin), touching anything without gloves, would conduct valuable body heat away, it was a tough trip. They ate three times as many Calories (to produce body heat) as normal and wore special clothing, which insulated against heat loss. Starting the trip with 100 tons of food and coal, and ending with only six people reaching the pole (the rest had turned back). It represented the end of the age of explorers; only climbing Mt. Everest remained for adventurers.

Since then a trip to the North Pole is much easier. People have flown planes, used snowmobiles, and even motorcycles to reach it. In fact modern nuclear powered submarines have made the trip with ease under the ice and even surfaced, breaking through it. Times have changed.
http://voices.nationalgeographic.com/2009/04/06/peary_and_the_north_pole/

Always remember and never forget

The energy in an object is thermal energy, this is the motion (or energy) of the molecules shaking or moving around. An object with a lot of thermal energy, cannot hurt you, even a red-hot branding iron is harmless, *as long as it is not touching you*. Thermal energy stays in the object. This chapter is about when that thermal energy *does move*; now that can hurt you. When thermal energy moves it is called heat, and heat can burn, because thermal energy is being added to YOU!

The other thing you need to remember is that there is no such thing as cold and therefore cold does not move between objects, cold cannot hurt you; *it does not even exist in science*. When you use the word "cold" in the real world, what you actually mean is there is a "lack of heat" or more properly, *the heat is moving away from you*, and your nerves tell your brain it is "cold". This is a survival mechanism to keep us safe.

The other thing to remember is that heat always moves from hot to cold, or high temperature to low temperature, it has nothing to do with which object has the most thermal energy. For example a cup of boiling hot water (with a tiny amount of thermal energy) placed in the cold ocean (with a lot of thermal energy), the heat moves from the hot water from the cup to the cold water, until the temperatures are the same.

When you put ice cubes in a glass of water the heat from the water moves into the ice, the ice does not make the water cold, the water makes the ice melt, but since the water loses energy, the water feels "colder" to us. This is confusing but actually makes sense.

Heat moves three ways

I like Gummy Bears and I like them cooked. There are three ways I can cook my gummy bears. I can put one on a spit above the fire, slowly rotating it, and let the rising heat cook it; I can put it to the side of the fire (like a marshmallow) and let the heat cook it that way until it is golden brown. I could also throw it right into the coals to cook. All of these work. Heat moves into my Gummy Bear.

HOW TO COOK A GUMMY BEAR

Heat that is rising is called convection (the top Gummy Bear), heat that cooks from the side is radiation (the one on the stick), and heat that cooks by touching the Gummy Bear is called conduction. All these methods will cook the Gummy Bear, and they are delicious.

1. Conduction (con "*touch*" tion)

This is when a hot object, like the coal in the fire *touches* a cooler object, like the gummy bear. The heat moves from the fire into the Gummy Bear and it gets hot enough to cook.

HOW TO COOK A GUMMY BEAR

Since the Gummy Bear is *touching* the hot coals, the heat goes right into the Gummy Bear. This is how I like to cook Potatoes in a campfire, just wrap them in aluminum foil and throw them in, corn on the cob is cooked the same at my house.

So **conduction** is when *heat moves between touching objects*. You experience this all the time when you touch something hot like the kitchen stove or a fire. It also happens when you put your warm tongue on a cold flagpole, the heat from you tongue goes into the flagpole and your tongue freezes to the pole (do not do this by the way and if you do, pour cold water on your tongue to melt the ice). A better way to explore conduction is to climb into a nice warm bathtub, the heat moves into you by conduction and you feel warm since the water is touching you.

Some objects allow heat to move through them very fast, these are called **conductors**, and are usually made of metal. Conductors move heat through them very fast, the word conductor means to "carry", like a train conductor carries passengers on the train. When you touch a *heat conductor*, it feels cold to you, because it sucks the heat out of your hand quickly. Touch a metal object on a cold day and you will realize this.

Some objects *prevent heat from moving*, these are called **insulators.** This is why you wear gloves and thick coats in the winter, the heat gets trapped and cannot move, the heat stays in you and you feel warm. We put insulation in the walls of our houses to keep the heat in the house (in the winter) and the heat out (in the summer). Insulators stop the movement of heat. They help us control the movement of heat.

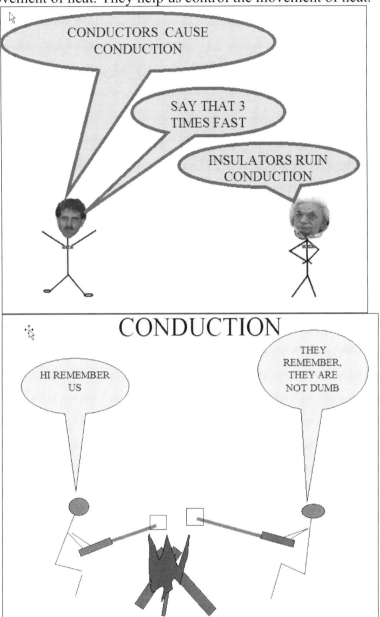

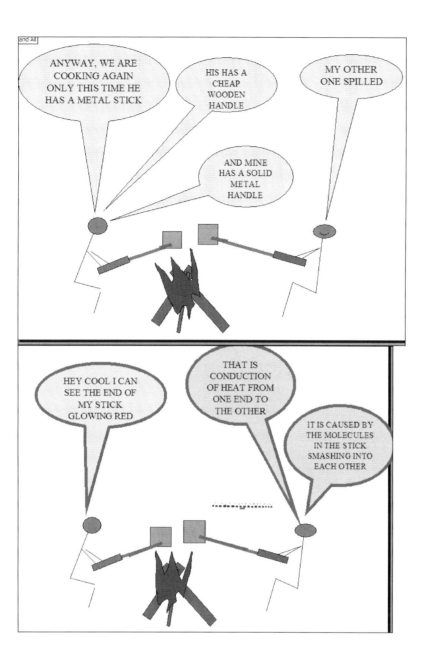

222

224

This is a good time to check out Eureka conduction - https://www.youtube.com/watch?v=Yitiw6Y7xZg

2. Convection

Due to density, hot air (fluffy) rises, while cold air (denser) sinks. The air above a fire rises because it is hotter than the surrounding air. Hot water, or any liquid, does the same. Hot molecules spread out and become less dense (or fluffy); the air above a fire rises and heats anything above it. This is another way I cook my Gummy Bear.

Convection is *when heat spreads by the movement of hot gas or liquid.* Hot fluids rise and cold fluids sink. Hot air balloons work best in the morning because the outside air is cooler; it pushes the hot air balloon upward. Hot dogs cook best above the fire due to convection.

Convection is why there are coastal winds at the beach. In the day time the hot sand heats the air above it and it rises, the cooler ocean makes the air cool and it sinks, the wind goes from cooler dense air above the ocean, to less dense, rising air over the sand. This is why kites fly inland at the beach; the wind always blows from the ocean.

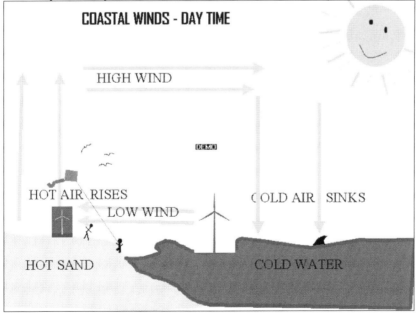

At night the opposite happens, and kites fly over the water.

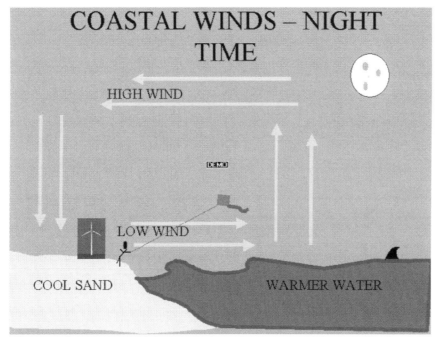

COASTAL WINDS – NIGHT TIME

HIGH WIND

DEMO

LOW WIND

COOL SAND   WARMER WATER

This is why the coast is a good place for wind turbines, it is always windy!

Check out this Eureka video on convection:

https://www.youtube.com/watch?v=NuxMRpWvsdw

3. Radiation

The other way heat moves is by shining on an object. This is another way to cook a Gummy Bear. The light from the fire *shines sideways* on my Gummy Bear and it absorbs the heat.

HOW TO COOK A GUMMY BEAR

The word radiation means *to shine,* all light can shine, heat is a form of light (electromagnetic energy), and it is called *infrared light*. Not only can Gummy Bears be cooked this way, so can marshmallows, in fact this is how we cook our marshmallows in my house. We make S'mores. **Radiation** is *when heat travels as light*. The heat from the sun reaches earth in this way; standing in front of a fire to get warm is radiation. The objects do not need to touch. Remember heat always moves from high temperature objects (like the sun) to low temperature objects like the Earth. Radiation can shine in any direction, including sideways.

It is kind of like when a child burns ants with a convex magnifying lens. The radiation is concentrated at one spot.

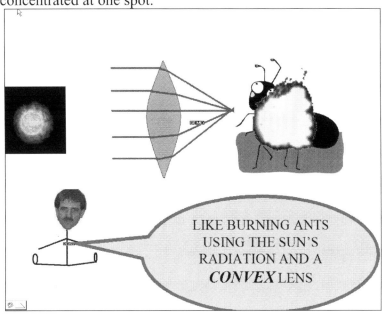

Back to our snake friends

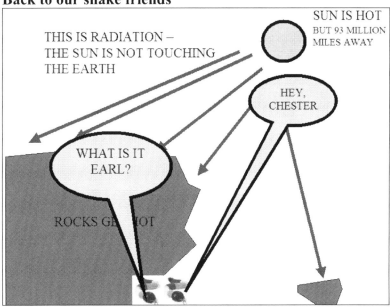

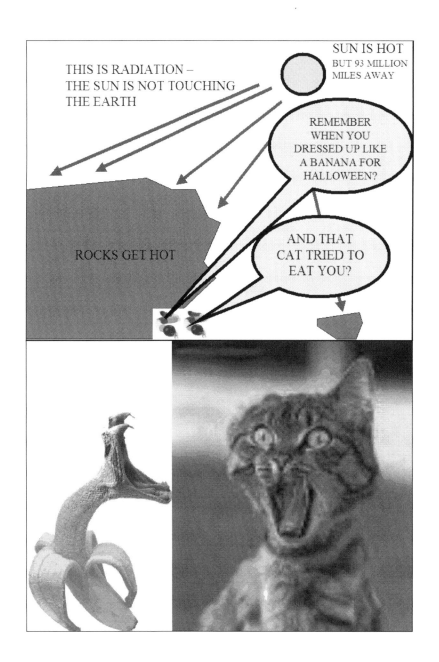

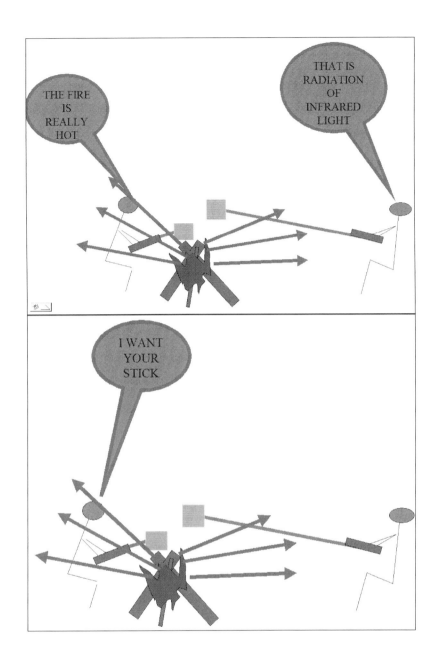

231

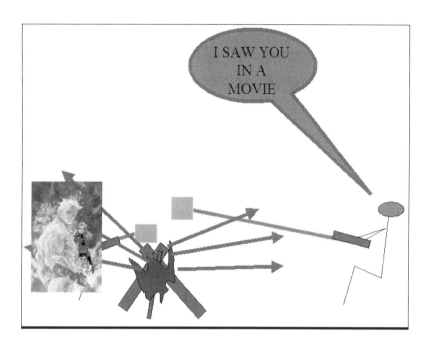

This is a good time to check out
Eureka – Radiation - https://www.youtube.com/watch?v=2JZciWtK6vc

233

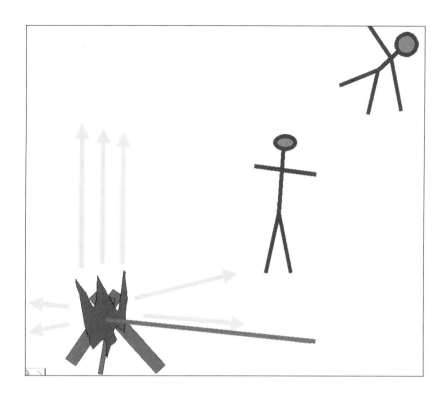

Keeping heat in

One of the goals of our society to is control heat movement, and the best way to do this is with insulation which slows the movement of heat. We put insulation in our houses to keep heat in, in the winter, or out, in the summer.

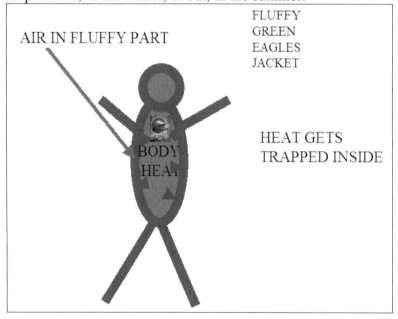

In review

CONDUCTION	CONVECTION	RADIATION
HEAT SPREADING ALONG A METAL POLE	HEAT RISING OVER A FIRE	SUNLIGHT
BURNING YOUR HAND WHEN YOU TOUCH A STOVE	HEAT SPREAD BY WATER MOLECULES	

Cats are a pain, at least mine can be a pain, and she is always in the perfect spot in the house. Cats know about heat, wherever your cat is, is where you want to be. Cats are smart and very selfish. They control us.

HOW IS THE CAT STAYING WARM??

CONVECTION FROM THE HEATER

IF THERE WAS A FIRE
IN THE ROOM IT WOULD BE

RADIATION AGAIN

A day in the life of Earwig Hickson III

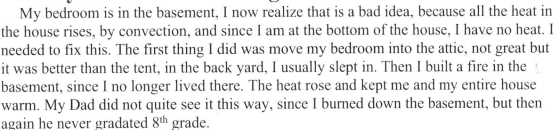

My bedroom is in the basement, I now realize that is a bad idea, because all the heat in the house rises, by convection, and since I am at the bottom of the house, I have no heat. I needed to fix this. The first thing I did was move my bedroom into the attic, not great but it was better than the tent, in the back yard, I usually slept in. Then I built a fire in the basement, since I no longer lived there. The heat rose and kept me and my entire house warm. My Dad did not quite see it this way, since I burned down the basement, but then again he never gradated 8th grade.

Once I had the attic bedroom I began to add a lot of insulation to the roof. I used the pink stuff from the home improvement store (until my Dad found out and took his credit card back). I added layers of newspaper, which was free and a decent insulation. I put anything "fluffy" I could find in the roof to stop the heat from leaving my new bedroom; this included all my stuffed animals and all my sister's ugly clothes. I looked up on the Internet how the early Pioneers insulated their houses and added Bull Rushes (a plant) to my insulation. All was well, until my dad came up one night and realized I had the warmest room in the house. Now I live back in the basement and he and Mom live in the attic. He not only says it is warm but extremely peaceful from annoying little brats, whatever that means.

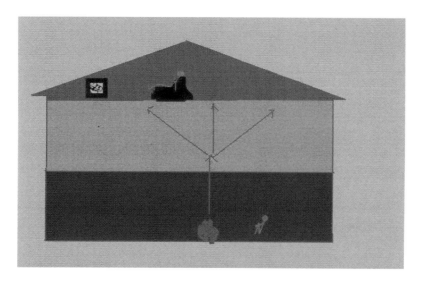

Review of terms – Quizlet:
https://Quizlet.com/124265761/chapter-18-volume-one-the-movement-of-heat-flash-cards/?new

Things to Google
Hot air balloons
Coastal winds
How tornadoes form
Fire tornado

Links
Hot and cold water convection – hot water rises and cold water sinks since hot liquid is less dense than cold liquid. https://www.youtube.com/watch?v=McZ9w7PdLqo

This is a fun one- I light my friend's head on fire, knowing that convection (hot air rising) will save him. https://www.youtube.com/watch?v=smknECV63uI

Here it is again, this was fun - https://www.youtube.com/watch?v=KvkLC-dSZYA

 Fire tornado – this works by convection of hot air (the green tornado) rising and the cold air around it sinking. The result is cool. The green is caused by Boric acid in the fire. https://www.youtube.com/watch?v=e7CashWyobw

The three ways heat moves with Hillary and Bill - https://www.youtube.com/watch?v=tLZNv2IoMiE

Two pieces of ice at the same temperature, but one melts faster. Magic or science? A film by Ted-ed. http://ed.ted.com/featured/qLn0UWUG

Conduction – two plates, one a conductor and one an insulator are placed on a table and an ice cube is placed on each – one ice cube will melt fast, which is it? https://www.youtube.com/watch?v=v9PHcxwu1Bw

Convection of water in flasks: https://www.youtube.com/watch?v=McZ9w7PdLqo

Convection of warm water: https://www.youtube.com/watch?v=4f5Fd7Ia-ys

This ends Volume one of this series.

Volume 2 – Motion, forces, and physics: https://www.amazon.com/dp/1520408021

Volume 3 – Chemistry, waves, and pseudoscience: Will be available by December 2019 on Amazon.

If you liked the stories about young Earwig Hickson III, you might like some of the stories of Earwig as an adult in the book. Method is Everything: A sportsman's Reflections and Misadventures, A Memoir, by N.Y. Best Sellers at: https://www.amazon.com/dp/1719815585

Note: The power point lessons are now available for this book on Google slides. E-mail: junglecat3388@gmail.com for details.